Impact of Carbon Dioxide, Trace Gases, and Climate Change on Global Agriculture

Impact of Carbon Dioxide, Trace Gases, and Climate Change on Global Agriculture

Proceedings of a symposium sponsored by Divisions A-3, A-6, S-6, A-5, C-3, S-1, C-2, S-3, and S-5 of the American Society of Agronomy, Crop Science Society of America, and Soil Science Society of America with support from the USDA-ARS. The symposium was held in Anaheim, CA, 1 December 1988.

Editorial Committee

Bruce A. Kimball, *chair*
Norman J. Rosenberg
Leon Hartwell Allen, Jr.

Editor-in-Chief ASA

G. H. Heichel

Editor-in-Chief CSSA

C. W. Stuber

Editor-in-Chief SSSA

D. E. Kissel

Managing Editor

Susan Ernst

ASA Special Publication Number 53

American Society of Agronomy, Inc.
Crop Science Society of America, Inc.
Soil Science Society of America, Inc.

1990

Cover Design: Patricia J. Scullion

American Society of Agronomy, Inc.
Crop Science Society of America, Inc.
Soil Science Society of America, Inc.
677 South Segoe Road, Madison, WI 53711 USA

Library of Congress Cataloging-in-Publication Data

Impact of carbon dioxide, trace gases, and climate change on global agriculture : proceedings of a symposium / sponsored by divisions A-3 . . . [et al.] of the American Society of Agronomy, Crop Science Society of America, and Soil Science Society of America with support from the USDA-ARS ; editorial committee, Bruce A. Kimball, chair . . . [et al.].
p. cm. — (ASA special publication ; no. 53)
"The symposium was held in Anaheim, CA, 1 December 1988."
Includes bibliographical references.
ISBN 0-89118-103-2 (paperback)
1. Climatic changes—Congresses. 2. Meteorology, Agricultural—Congresses. 3. Atmospheric carbon dioxide—Congresses. 4. Greenhouse gases—Congresses. I. Kimball, Bruce A. II. American Society of Agronomy. Division A-3. III. Crop Science Society of America. IV. Soil Science Society of America. V. Series.
S1.A453 no. 53
[S600.7.C64]
630 s—dc20
[632′.1] 90-37086
CIP

Printed in the United States of America

CONTENTS

FOREWORD

Global climate change is one of several important issues that will command the attention of policymakers and scientists in the 1990s. The evidence that concentrations of carbon dioxide (CO_2), and other gases are increasing in the atmosphere is irrefutable. This evidence, and the knowledge that CO_2 and trace gases may absorb thermal radiation sufficient to warm the atmosphere, has prompted much speculation that ensuing atmospheric warming may lead to changes in the distribution of precipitation, and of crop adaptation and productivity, that would alter the world supply of food and fiber. The implications of this speculation are compelling for agronomists, because agronomists are stewards of the world's food supply and of the natural resources that are used to produce food. Agronomists have a pivitol role in conducting the research needed to anticipate crop response to climate change, and in informing policymakers and the general public about the adequacy of our knowledge.

In this publication, leading scientists from the international community of agronomists assess the current status of scientific knowledge about the putative role of greenhouse gases in global climate change and report their findings. The information will be useful not only to scientists, but also to policymakers as background for deliberation on farm, environmental, and agricultural research legislation.

A.A. BALTENSPERGER, *president*
American Society of Agronomy

STEVE A. EBERHART, *president*
Crop Science Society of America

WILFORD R. GARDNER, *president*
Soil Science Society of America

GARY H. HEICHEL, *editor-in-chief*
ASA Publications

PREFACE[1]

The concentrations of CO_2 and some other radiatively active trace gases are increasing in the global atmosphere, and these changes in composition have been predicted to significantly warm the Earth (the "greenhouse effect") and possibly to alter precipitation patterns, as described by Taylor and MacCracken (Chapter 1). Agricultural productivity is greatly dependent on climate and weather, so these predictions or projections of climatic change should be of concern to agronomists and other agricultural scientists. Moreover, Burke and Lashof (Chapter 3) indicate agricultural practices and changing patterns of land use affect the emission rates of greenhouse gases. So to some extent, agriculture not only is vulnerable to the effects of climate change, but it is a possible cause as well. Not everyone agrees with the prediction of warming (Idso, Chapter 2), but nevertheless, it is prudent that we begin to consider the possibility of climate change and to make our planning for the future of agriculture as flexible as possible.

Even if climate does not change, the increased CO_2 concentration will significantly affect the growth of plants and probably increase agricultural productivity (Acock, Chapter 4). If climate does change, however, significant interactions are likely to occur between the CO_2 and the climate variables in their effects on plant growth (Idso, Chapter 5; Goudriaan & Unsworth, Chapter 8). These growth changes will not only affect our crop plants, but they will also affect the weeds with which they compete; and of course, competitive advantages among species growing in the wild may shift, so that the composition of plant communities may be altered (Patterson & Flint, Chapter 7). If the climate warms, soil organic matter content is likely to decrease while total N may increase in the temperate zone, but no major changes in fertilization practices are predicted (Buol et al., Chapter 6). Yet, even while global CO_2 and climate changes may be affecting worldwide agricultural production, the availability of one important resource for coping with hotter and drier conditions, irrigation water, may be adversely affected (Goudriaan & Unsworth, Chapter 8). Thus, in this book we attempt to review what can reasonably be inferred from the scientific literature about the effects of elevated CO_2 and changing climate on crop production and to describe their possible consequences for the future of agriculture.

This book is a direct result of a symposium held on 1 December 1988 at the American Society of Agronomy Annual Meeting in Anaheim, CA. Each chapter has been written by an invited expert speaker to that symposium, with the help of coauthors they have enlisted. Early in 1987, when first elected to be the 1988 program chair for Division A-3, several colleagues urged me to organize a symposium on this timely topic. That it was timely proved to be an understatement when the Midwest drought of the summer of 1988 became headline news across the USA and raised the public consciousness level about the possibility that the greenhouse effect might be real and that the climate might already have changed. Only a few atmospheric scientists publicly attributed the USA drought of 1988 to the greenhouse effect.

[1] Contribution from the USDA-ARS.

Nevertheless, the interest in the topic and the symposium was high, and hopefully will also be high for this book.

In addition to the invited speakers (who, along with their coauthors, have produced the chapters in this book), the symposium also featured a session with 14 volunteered posters. Although the goal of the invited papers and this book was to review existing information and synthesize principles, the posters presented new research results, which also contributed to the interest in and success of the symposium. The abstracts of these poster presentations are in the 1988 Agronomy Abstracts (published by the American Society of Agronomy), but as an acknowledgement of their important contribution to the symposium and as an aid to the reader, a list of the poster presentations is included in Appendix A.

Several other acknowledgments are appropriate for contributions to the symposium and to this book. First, I thank the speakers and presenters who produced stimulating talks and posters that made the symposium a success. Similarly, I thank the authors and coauthors for writing their book chapters that provide an excellent compilation of current knowledge on this topic. Also toward this end, I appreciate the efforts of my colleagues on the editorial committee, Drs. Norman J. Rosenberg and Leon Hartwell Allen, Jr., who took their role seriously by rigorously and fairly reviewing and editing the manuscripts. Next, I wish to thank the 1988 program chairs of the several other divisions who cosponsored the symposium, particularly those who pledged financial support. The encouragement and the solicitations of financial support for the symposium by 1988 ASA program chair Dr. E.C.A. Runge are greatly appreciated. Similarly, the encouragement and suggestions of Dr. Roger C. Dahlman, Department of Energy, Carbon Dioxide Research Division, were helpful in shaping the symposium. Finally, the grant from the USDA-ARS, which provided travel expenses for the "non-society" speakers, is gratefully acknowledged.

BRUCE A. KIMBALL, *editor*
U.S. Water Conservation Laboratory, Phoenix, Arizona

CONTRIBUTORS

Basil Acock	Research Leader, USDA-ARS, NRI, Systems Research Laboratory, Beltsville, MD 20705
Leon Hartwell Allen, Jr.	Soil Scientist, USDA-ARS, University of Florida, Gainesville, FL 32611
S. W. Buol	Professor of Soil Science, Soil Science Department, North Carolina State University, Raleigh, NC 27695-7619
Lauretta M. Burke	Environmental Analyst, AMS, Arlington, VA 22009. Formerly Environmental Analyst, The Bruce Company, Washington, DC
Elizabeth P. Flint	Research Associate, Department of Botany, Duke University, Durham, NC 27708
J. Goudriaan	Crop Physiologist, Department of Theoretical Production Ecology, Agricultural University, Wageningen, the Netherlands
Sherwood B. Idso	Research Physicist, USDA-ARS, U.S. Water Conservation Laboratory, Phoenix, AZ 85040
Bruce A. Kimball	Soil Scientist, USDA-ARS, U.S. Water Conservation Laboratory, Phoenix, AZ 85040
J. M. Kimble	Research Soil Scientist, USDA-SCS, Lincoln, NE 68508-3866
Daniel A. Lashof	Senior Scientist, Natural Resources Defense Council, Washington, DC 20005. Formerly Environmental Scientist, USEPA, Washington, DC
Michael C. MacCracken	Division Leader, Atmospheric and Geophysical Sciences, Lawrence Livermore National Laboratory, Livermore, CA 94550
David T. Patterson	Plant Physiologist, USDA-ARS, Botany Department, Duke University, Durham, NC 27706
Norman J. Rosenberg	Senior Fellow and Director, Climate Resources Program, Resources for the Future, Washington, DC 20036
Pedro A. Sanchez	Professor of Soil Science, North Carolina State University, Raleigh, NC 27695-7619
Karl E. Taylor	Research Scientist, Lawrence Livermore National Laboratory, Livermore, CA 94550
M. H. Unsworth	Department of Physiology and Environmental Science, University of Nottingham, School of Agriculture, Sutton Bonington, LE12 5RD, England
S. B. Weed	Professor of Soil Science, Department of Soil Science, North Carolina State University, Raleigh, NC 27695-7619

Conversion Factors for SI and non-SI Units

Conversion Factors for SI and non-SI Units

To convert Column 1 into Column 2, multiply by	Column 1 SI Unit	Column 2 non-SI Unit	To convert Column 2 into Column 1, multiply by
		Length	
0.621	kilometer, km (10^3 m)	mile, mi	1.609
1.094	meter, m	yard, yd	0.914
3.28	meter, m	foot, ft	0.304
1.0	micrometer, μm (10^{-6} m)	micron, μ	1.0
3.94×10^{-2}	millimeter, mm (10^{-3} m)	inch, in	25.4
10	nanometer, nm (10^{-9} m)	Angstrom, Å	0.1
		Area	
2.47	hectare, ha	acre	0.405
247	square kilometer, km^2 $(10^3\ m)^2$	acre	4.05×10^{-3}
0.386	square kilometer, km^2 $(10^3\ m)^2$	square mile, mi^2	2.590
2.47×10^{-4}	square meter, m^2	acre	4.05×10^3
10.76	square meter, m^2	square foot, ft^2	9.29×10^{-2}
1.55×10^{-3}	square millimeter, mm^2 $(10^{-3}\ m)^2$	square inch, in^2	645
		Volume	
9.73×10^{-3}	cubic meter, m^3	acre-inch	102.8
35.3	cubic meter, m^3	cubic foot, ft^3	2.83×10^{-2}
6.10×10^4	cubic meter, m^3	cubic inch, in^3	1.64×10^{-5}
2.84×10^{-2}	liter, L (10^{-3} m^3)	bushel, bu	35.24
1.057	liter, L (10^{-3} m^3)	quart (liquid), qt	0.946
3.53×10^{-2}	liter, L (10^{-3} m^3)	cubic foot, ft^3	28.3
0.265	liter, L (10^{-3} m^3)	gallon	3.78
33.78	liter, L (10^{-3} m^3)	ounce (fluid), oz	2.96×10^{-2}
2.11	liter, L (10^{-3} m^3)	pint (fluid), pt	0.473

		Mass	
2.20×10^{-3}	gram, g (10^{-3} kg)	pound, lb	454
3.52×10^{-2}	gram, g (10^{-3} kg)	ounce (avdp), oz	28.4
2.205	kilogram, kg	pound, lb	0.454
0.01	kilogram, kg	quintal (metric), q	100
1.10×10^{-3}	kilogram, kg	ton (2000 lb), ton	907
1.102	megagram, Mg (tonne)	ton (U.S.), ton	0.907
1.102	tonne, t	ton (U.S.), ton	0.907
		Yield and Rate	
0.893	kilogram per hectare, kg ha^{-1}	pound per acre, lb acre^{-1}	1.12
7.77×10^{-2}	kilogram per cubic meter, kg m^{-3}	pound per bushel, bu^{-1}	12.87
1.49×10^{-2}	kilogram per hectare, kg ha^{-1}	bushel per acre, 60 lb	67.19
1.59×10^{-2}	kilogram per hectare, kg ha^{-1}	bushel per acre, 56 lb	62.71
1.86×10^{-2}	kilogram per hectare, kg ha^{-1}	bushel per acre, 48 lb	53.75
0.107	liter per hectare, L ha^{-1}	gallon per acre	9.35
893	tonnes per hectare, t ha^{-1}	pound per acre, lb acre^{-1}	1.12×10^{-3}
893	megagram per hectare, Mg ha^{-1}	pound per acre, lb acre^{-1}	1.12×10^{-3}
0.446	megagram per hectare, Mg ha^{-1}	ton (2000 lb) per acre, ton acre^{-1}	2.24
2.24	meter per second, m s^{-1}	mile per hour	0.447
		Specific Surface	
10	square meter per kilogram, m^2 kg^{-1}	square centimeter per gram, cm^2 g^{-1}	0.1
1000	square meter per kilogram, m^2 kg^{-1}	square millimeter per gram, mm^2 g^{-1}	0.001
		Pressure	
9.90	megapascal, MPa (10^6 Pa)	atmosphere	0.101
10	megapascal, MPa (10^6 Pa)	bar	0.1
1.00	megagram per cubic meter, Mg m^{-3}	gram per cubic centimeter, g cm^{-3}	1.00
2.09×10^{-2}	pascal, Pa	pound per square foot, lb ft^{-2}	47.9
1.45×10^{-4}	pascal, Pa	pound per square inch, lb in^{-2}	6.90×10^3

(continued on next page)

Conversion Factors for SI and non-SI Units

To convert Column 1 into Column 2, multiply by	Column 1 SI Unit	Column 2 non-SI Unit	To convert Column 2 into Column 1, multiply by
	Temperature		
1.00 (K − 273)	Kelvin, K	Celsius, °C	1.00 (°C + 273)
(9/5 °C) + 32	Celsius, °C	Fahrenheit, °F	5/9 (°F − 32)
	Energy, Work, Quantity of Heat		
9.52×10^{-4}	joule, J	British thermal unit, Btu	1.05×10^{3}
0.239	joule, J	calorie, cal	4.19
10^{7}	joule, J	erg	10^{-7}
0.735	joule, J	foot-pound	1.36
2.387×10^{-5}	joule per square meter, J m^{-2}	calorie per square centimeter (langley)	4.19×10^{4}
10^{5}	newton, N	dyne	10^{-5}
1.43×10^{-3}	watt per square meter, W m^{-2}	calorie per square centimeter minute (irradiance), cal cm^{-2} min^{-1}	698
	Transpiration and Photosynthesis		
3.60×10^{-2}	milligram per square meter second, mg m^{-2} s^{-1}	gram per square decimeter hour, g dm^{-2} h^{-1}	27.8
5.56×10^{-3}	milligram (H_2O) per square meter second, mg m^{-2} s^{-1}	micromole (H_2O) per square centimeter second, μmol cm^{-2} s^{-1}	180
10^{-4}	milligram per square meter second, mg m^{-2} s^{-1}	milligram per square centimeter second, mg cm^{-2} s^{-1}	10^{4}
35.97	milligram per square meter second, mg m^{-2} s^{-1}	milligram per square decimeter hour, mg dm^{-2} h^{-1}	2.78×10^{-2}
	Plane Angle		
57.3	radian, rad	degrees (angle), °	1.75×10^{-2}

	Electrical Conductivity, Electricity, and Magnetism		
10	siemen per meter, S m^{-1}	millimho per centimeter, mmho cm^{-1}	0.1
10^4	tesla, T	gauss, G	10^{-4}
	Water Measurement		
9.73×10^{-3}	cubic meter, m^3	acre-inches, acre-in	102.8
9.81×10^{-3}	cubic meter per hour, $m^3\ h^{-1}$	cubic feet per second, $ft^3\ s^{-1}$	101.9
4.40	cubic meter per hour, $m^3\ h^{-1}$	U.S. gallons per minute, gal min^{-1}	0.227
8.11	hectare-meters, ha-m	acre-feet, acre-ft	0.123
97.28	hectare-meters, ha-m	acre-inches, acre-in	1.03×10^{-2}
8.1×10^{-2}	hectare-centimeters, ha-cm	acre-feet, acre-ft	12.33
	Concentrations		
1	centimole per kilogram, cmol kg^{-1} (ion exchange capacity)	milliequivalents per 100 grams, meq 100 g^{-1}	1
0.1	gram per kilogram, g kg^{-1}	percent, %	10
1	milligram per kilogram, mg kg^{-1}	parts per million, ppm	1
	Radioactivity		
2.7×10^{-11}	becquerel, Bq	curie, Ci	3.7×10^{10}
2.7×10^{-2}	becquerel per kilogram, Bq kg^{-1}	picocurie per gram, pCi g^{-1}	37
100	gray, Gy (absorbed dose)	rad, rd	0.01
100	sievert, Sv (equivalent dose)	rem (roentgen equivalent man)	0.01
	Plant Nutrient Conversion		
	Elemental	***Oxide***	
2.29	P	P_2O_5	0.437
1.20	K	K_2O	0.830
1.39	Ca	CaO	0.715
1.66	Mg	MgO	0.602

1 Projected Effects of Increasing Concentrations of Carbon Dioxide and Trace Gases on Climate[1]

Karl E. Taylor and Michael C. MacCracken

Lawrence Livermore National Laboratory
Livermore, California

ABSTRACT

The atmospheric concentrations of CO_2, CH_4, N_2O, chlorofluorocarbons, and other gases capable of trapping terrestrial infrared radiation have risen markedly since preindustrial times. Theoretical and numerical analyses agree that these changes will cause a warming of the global average temperature by perhaps 1.5 to 4.5 °C during the 21st century, with actual changes dependent on rates of emissions. Evaporation and precipitation rates will also increase, but there is no consensus as to what specific changes will occur in different regions. Although many uncertainties remain, changes in temperature and soil water availability seem likely to induce changes in agricultural and natural ecological systems, with the rate, pattern, and magnitude of the changes being the most important—but still uncertain—determinants.

Historically, farmers have tuned their agricultural practices to the natural variations of the local climate. As climate changed, societies have had little choice but to adapt or migrate. Until the 20th century most regional climate change was a manifestation of natural variations of the global climate system or was more directly caused by natural processes acting locally. The agricultural and societal disruption caused by climate change was considered a natural and inherent consequence of living on planet Earth. In the 1980s, however, there has been rising concern that over the next several decades natural variations in the climate system will no longer dominate, but will be swamped by global climate changes brought on by societal activities. In particular, we

[1] This work was performed under the auspices of the Department of Energy Carbon Dioxide Research Program by the Lawrence Livermore National Laboratory under contract W-7405-ENG-48.

know that anthropogenic emissions of various gases have begun to alter the composition of the atmosphere, primarily through combustion of fossil fuels, manufacture and release of various chemicals, and changes in land use. The effects of these alterations in atmospheric composition will include modification of the climate to which society has become accustomed. Because it is expected that these anthropogenically induced changes in climate will grow to be more persistent and extensive than the natural climate variations experienced in the preceding several thousand years, and because these changes could occur relatively rapidly over a period of < 100 yr, scientists and others have become concerned that adapting to climate change will prove more difficult than in the past.

Humanity may now be altering global climate by increasing the atmospheric concentrations of certain trace gases such as CO_2, CH_4, N_2O, chlorofluorocarbons, and O_3, which contribute to what is known as the *atmospheric greenhouse effect. Greenhouse gases* are essentially transparent to solar radiation, but absorb and reradiate downward a significant fraction of the infrared radiation emitted by the Earth's surface and gases in the lower atmosphere. This interference with the transmission of the infrared radiation to space leads to a warming of climate in a manner roughly analogous to the way in which a greenhouse warms as its glass walls allow sunlight to penetrate, but prevent the warmed air from exiting. The greenhouse analogy is defective, but the term *greenhouse effect* is nevertheless nearly universally accepted.

The importance of greenhouse gases is clearly demonstrated by considering their effect on the radiation balance at the top of the atmosphere. If all greenhouse gases were suddenly moved from the atmosphere (and the greenhouse effect of clouds were also eliminated), we would observe an immediate increase in the radiative cooling rate of the Earth of about 150 W/m^2 (approximately a 60% increase). In the parlance of climate modelers, the greenhouse effect provides a "radiative forcing" of the climate system of 150 W/m^2. For comparison, the net solar radiative forcing is approximately 240 W/m^2 (given that roughly 30% of the solar insolation is reflected by the Earth and its atmosphere). Because the radiative processes are well understood, there is no doubt that the atmospheric greenhouse effect is real. Energy trapped in the lower atmosphere by clouds, water vapor, and other greenhouse gases maintains the average surface temperature of the Earth some 33 K warmer than it would otherwise be. Still, much uncertainty remains concerning the magnitude, pattern, and timing of current and future temperature increases and other climate changes that will result from the changes in atmospheric composition caused by humans.

This chapter will provide a summary of some of the scientific issues that comprise the trace gas–climate change problem. More detailed reviews and further discussion of the subject can be found in Schneider (1989), Ramanathan (1988), Ramanathan et al. (1987), Wang et al. (1986), Dickinson and Cicerone (1986), Schlesinger and Mitchell (1987), and a series of reports edited by MacCracken and Luther (1985a, b) and Trabalka (1985). This chapter will focus on CO_2 and its effects on temperature. The next section begins by describing the global carbon cycle and the changes that have occurred since the 1860s. A discussion follows of how climate models are used to determine the climatic response to increasing concentrations of greenhouse gases. The uncertainties in these model projections are emphasized. In the following section, the record of climate observations is discussed and its use, both for validating models and for verifying their projections, is described. In the concluding section, we consider some of the consequences of the projected climate change for natural and managed ecosystems.

THE CARBON CYCLE

To appreciate society's potential for causing significant changes in atmospheric composition, it is useful to consider the primary features of the global carbon cycle. Trabalka (1985) provides a more complete summary. Carbon resides on Earth in several different forms and the carbon cycle is characterized by exchanges over a wide range of time-scales between reservoirs. The important reservoirs for carbon that can contribute to changes in atmospheric CO_2 on time scales of <100 yr include the biosphere, the oceans, and fossil fuel reserves. There are about 740 Pg (1 Pg = 1 petagram = 10^{15} g = 1 gigaton) of carbon in the atmosphere in the form of CO_2, accounting for the present atmospheric concentration of about 350 μmol/mol. A comparable amount of carbon resides in the active part of the biosphere, primarily in land plants. The mixed layer of the ocean (that is the upper 75 m or so that interact readily with the atmosphere) contains about 630 Pg. These amounts are dwarfed by the carbon that could be relcased if all known fossil fuels (3900 Pg) were combusted or if a sizeable fraction of the carbon stored in the deeper layers of the ocean (38 000 Pg) were somehow to be released to the atmosphere.

Before societal activities began to disrupt the carbon cycle, the exchange of carbon among these reservoirs was probably in near balance (that is, major shifts occurred only over periods of thousands of years or more). Under these conditions, the release of 130 Pg/yr of carbon (as CO_2) by dead plant matter was nearly balanced by uptake of carbon by new plant growth. Similarly, about 160 Pg of carbon were exchanged between the oceans and atmosphere each year, again with no net transfer of carbon. Today, however, combustion of fossil fuels is introducing about 5 Pg of carbon into the atmosphere each year. It has also been variously estimated that between 0 and 2.6 Pg/yr of additional carbon enters the atmosphere each year as a result of changes in land-use practices. A major concern in this regard is the destruction of vast expanses of tropical rain forests, which disrupts the balance between growing and decaying plant material. Overall, the activities of humankind are introducing about 6 ± 1 Pg of carbon/yr into the atmosphere. Measurements confirm that atmospheric concentrations of CO_2 are indeed increasing, but by only about half the rate of emissions. A large portion of the "missing" carbon is probably taken up by the oceans, although there is still some uncertainty about whether midlatitude forest regrowth or other fac-

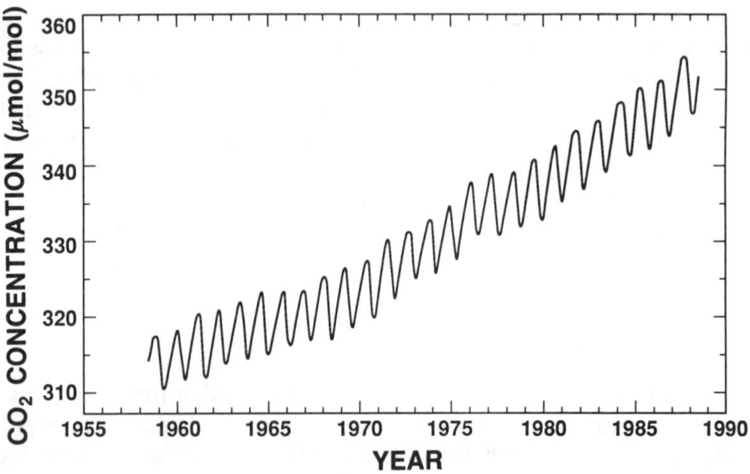

Fig. 1–1. Observations of atmospheric CO_2 concentration at the Mauna Loa Observatory in Hawaii vs. time. The trend reflects the gradual global increase in the annual average CO_2 concentration due mainly to fossil fuel emissions. The obvious seasonal variation is a result of the seasonal biological cycle with the spurt of new plant growth each spring exceeding the decay of old vegetation, thereby temporarily reducing atmospheric concentration of CO_2 until the end of seasonal plant growth in the fall. (Data from Keeling, 1989, personal communication, provided by the Carbon Dioxide Information Analysis Center, Oak Ridge National Laboratory, Oak Ridge, TN.)

tors may be playing a significant role. Scientists are also concerned that the ability of the oceans to remove CO_2 will decrease in the future due to warming of the upper ocean and reduced carbon transport to the deep ocean.

Even with the many uncertainties about the details of the carbon cycle, there is no question that the CO_2 concentration in the atmosphere continues to increase and that much—probably most—of this rise is a result of fossil fuel combustion, a result that can be confirmed by measuring the dilution of atmospheric ^{14}C that occurs as fossil fuel carbon (with no ^{14}C) is mixed into the atmosphere (Trabalka, 1985). Figure 1–1 shows that the CO_2 concentration in the atmosphere has increased by about 9% since about 1960. The cycle evident in this record results from the seasonal cycle of growth and decay of vegetation in the Northern Hemisphere.

It is likely that the trend shown in Fig. 1–1 will continue at a similar or perhaps accelerated pace in the future. Since the end of World War II, annual emissions have grown by a factor greater than three. Growth rates have been especially high in the People's Republic of China and developing countries in South Asia, Africa, and Latin America. By 1980 these countries accounted for >20% of the emissions, whereas in 1950 they accounted for about 7% (Rotty & Masters, 1985).

Carbon dioxide is not the only trace gas causing concern (see Burke & Lashof, Chapter 3, this book). The concentrations of several other radiatively active trace gases that contribute to the greenhouse effect are also rising (Ramanathan et al., 1987), as summarized by Table 1–1. Here, too, it appears that human activities have perturbed natural biogeochemical balances. In

Table 1-1. Preindustrial and current atmospheric concentrations of selected greenhouse gases, current growth rates, and equivalent CO_2 growth rate [compiled from Wuebbles & Edmonds (1988), Ramanathan et al. (1985), and Hansen et al. (1988)].

Chemical species	Preindustrial level	1980s level	Current growth rate	Equivalent CO_2 growth rate§
	μmol/mol		%/yr	
Carbon dioxide (CO_2)	~270†	~345	0.4	0.4
Methane (CH_4)	0.7	1.7	1.1	0.15
Nitrous oxide (N_2O)	0.28	0.30	0.3	0.05
Chlorocarbons				
CFC-11	0	0.000 23	4‡	0.035
CFC-12	0	0.000 39	4‡	0.07
Other CFCs, O_3, and strat. H_2O	--	--	--	0.10

† Value given is estimate for mid-19th century.
‡ Prior to implementation of chlorocarbon production limitations called for in the Montreal Protocol signed in 1988.
§ The equivalent CO_2 growth rate is defined to be the percent change per year of CO_2 concentration that would have the same impact on equilibrium surface temperature as the current growth rate of the given chemical species. The values are approximately accurate for present conditions only.

some cases, entirely new manmade gases such as the chlorofluorocarbons have entered the atmosphere. The potential consequences of these emissions have been the subject of intense study by climatologists, and the results of their research will be described in the following sections.

CLIMATE MODELS AND PROJECTIONS OF CLIMATE CHANGES

Several different methods could, in principle, be used to project future climate change induced by increases in greenhouse gases. Extrapolation from current trends is unacceptable, partly because natural climate fluctuations cannot be removed from the historical climate record. In addition, nonlinearities in the climate system preclude linear extrapolation, and the nature of the nonlinearities cannot be determined from the historical record. Attempts have been made to determine the sensitivity of the climate system empirically by examining the changes in radiative fluxes and temperatures that occur during the normal cycle of seasonal change and as a result of small-scale perturbations. These approaches have not proven successful because the time and space domains of these analyses have not been comparable to those of the CO_2 problem. A third method has been tried in which climate analogs to the current situation are sought. This involves a study of historical and paleoclimatic records in an attempt to find a past situation during which time similar or comparable changes were taking place. Because of the limitations in data bases, incomplete understanding of the causes of past climate changes, and an unprecedented rate of change of atmospheric CO_2 and trace gas concentrations, no perfectly suitable analog is available.

Because other methods have generally proven completely unsuccessful, we are forced to rely on climate models, which are also imperfect, but nevertheless can provide projections of climate change that are consistent with the natural laws that govern the climate system. *Climate models* are theoretical constructs based on the fundamental equations governing the motions of the atmosphere and flow of energy through the climate system. A hierarchy of models has been developed to study climate change. Among the more highly simplified models, *energy balance climate models* are designed to represent the major climate zones of the Earth, but because they are longitudinally and vertically averaged, they contain highly parameterized representations of the atmospheric circulation and vertical energy transport. *Radiative-convective models* are used to study the vertical structure of the atmosphere but provide no information about the horizontal variations of climate. *Two-dimensional zonally averaged climate models* have been developed that explicitly simulate the atmospheric structure as a function of latitude and altitude (but not longitude). Because at least one spatial dimension is eliminated in each of these models, simulations can be carried out with modest computational expenses. For this and other reasons, much of the exploratory work concerning the effects of changing atmospheric composition on climate has been performed using these simplified models.

To simulate the atmospheric circulation explicitly, however, three-dimensional models are obviously required, and the most comprehensive models of this type are *global general circulation models* (GCMs). Because these models contain fewer approximations and have higher horizontal resolution than the more simplified models, more confidence is placed in their projections. These models have much in common with global weather prediction models, but the climate models typically include more sophisticated formulations for clouds and for certain processes involving the oceans, sea-ice, and land surfaces.

Figure 1–2 shows the major components of a typical climate model. Equations representing the conservation laws for mass, momentum, energy, and water vapor are solved for the atmosphere and, in newer models, also for the oceans. Although processes such as radiation can be treated explicitly, some processes in the atmosphere must be treated parametrically because they are not well enough understood and/or because computers are not yet powerful enough to adequately treat the processes in sufficient detail. Clouds and small-scale convection are two important elements in climate models that can presently be represented only crudely through approximate parameterizations. There are several quantities that are prescribed in climate models. Among these are physical constants such as the rotation rate of the Earth and the acceleration of gravity, and measured quantities such as the insolation at the top of the atmosphere. Models of the carbon cycle are not generally used to drive the climate models. Atmospheric composition is therefore prescribed (sometimes in a time varying manner). To study the effects of greenhouse gases, the composition can be varied to calculate the climate response to various perturbations (e.g., emissions scenarios).

The atmosphere interacts with the Earth's surface—ocean, land, ice, or snow. Figure 1–2 shows that climate models typically account for conservation of energy, water, snow and sea-ice mass, although many important processes must be parameterized. The surface exchanges energy, momentum, and water vapor with the atmosphere above.

In general circulation climate models, the global domain of the atmosphere and oceans is subdivided into several thousands of grid-cells. Current practical constraints on computer resources typically limit the horizontal spatial resolution in climate models to several hundred kilometers. A single grid cell in California, for example, may represent the disparate climates of the coastal region, the Central Valley, Sierra Nevada mountains, and the Mojave Desert. Even if current climate models were perfectly accurate, their relatively coarse resolution would limit the usefulness of their projections on regional scales, because it is unlikely that the different climates within millions of hectares will change in a common way.

In recognition of these and many other limitations, climate models are being carefully tested to determine the extent to which their simulations are valid. Comparisons of radiation calculations in the models with results of more detailed radiative transfer models and with laboratory and field experiments indicate that this component of the climate models is quite accurately represented. The dynamical behavior of the atmosphere has been examined in weather forecast experiments, and the representation of large-scale dynamics is thought to be well done. To test the overall behavior of the models, they have been used to simulate the climate changes associated with the comparatively large changes in radiative forcing of the normal seasonal cycle.

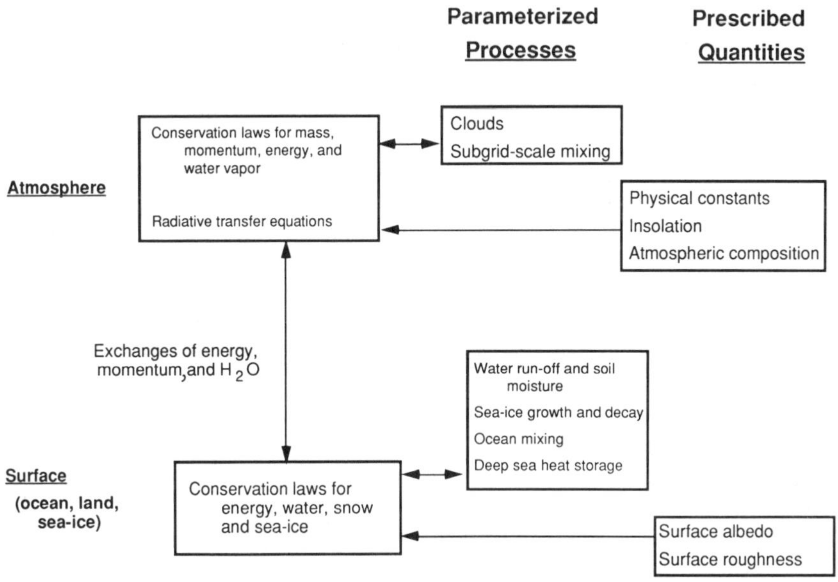

Fig. 1–2. Schematic diagram of the processes and interactions that are simulated in a typical climate model.

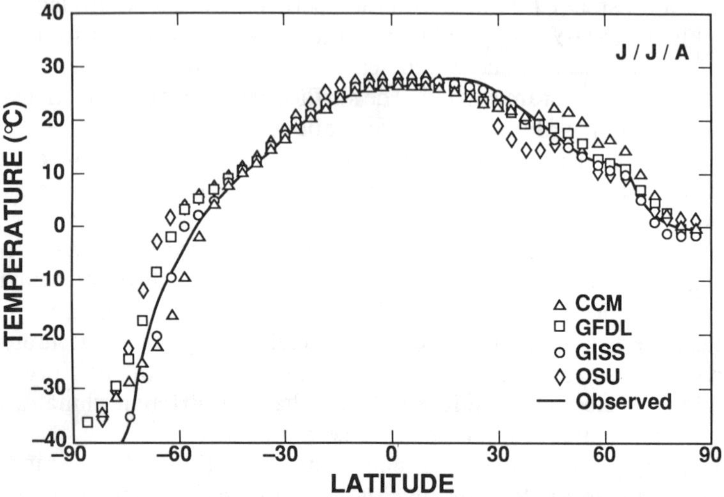

Fig. 1-3. Observed and model-computed median zonal surface air temperature, averaged over June, July, and August, vs. latitude (adapted from Grotch, 1988). [*CCM*—National Center for Atmospheric Research Community Climate model (Washington & Meehl, 1984); *GFDL*—Geophysical Fluid Dynamics Laboratory model (Wetherald & Manabe, 1986); *GISS*—Goddard Institute for Space Sciences model (Hansen et al., 1984); *OSU*—Oregon State University model (Schlesinger & Zhao, 1989); observations (Jenne, 1975)].

As an example of the ability of current climate models to simulate contemporary climatic conditions, Fig. 1-3 shows the zonal surface temperature calculated by four different climate models averaged over the months of June through August. The solid line connects the observations of surface air temperature. This and similar figures for the other seasons, reported by Grotch (1988), show that these models can reproduce the zonal character of climate as well as the ranges of seasonal average temperature (which are of the order of 20 K outside of the tropics). A closer look, however, reveals that there are cases where the models differ from observations by as much as 10 K, especially in regions where seasonal sea-ice is found. In general, models constructed by different research groups (which differ in their detailed representations of various atmospheric and surface processes) are in broad agreement with each other and are consistent with observations when average conditions over large spatial scales are considered, but important differences are found at continental and smaller scales. For example, Table 1-2 shows that over the western USA, four different climate models developed at U.S. research centers are able to simulate the observed temperature within a few degrees in most cases, but individual grid points may differ from observations by as much as 12 K. Much current research is being undertaken to determine why the differences arise in seemingly similar models. This should eventually lead to improvements in the models.

Climate models have been under development since the late 1960s and through extensive analysis scientists have slowly gained confidence in their ability to simulate climate responses to certain kinds of perturbations. The

Table 1-2. Control temperatures (°C) as represented by four general circulation models and derived from an observational data set for a roughly comparable set of grid points over the western USA.

Model mnemonic	Number of grid points	Minimum value	Maximum value	Median value
		December–January–February		
CCM†	6	−3.2	10.3	3.3
GFDL‡	6	−8.1	10.3	1.7
GISS§	4	−6.9	10.1	0.0
OSU¶	8	−5.9	7.4	1.2
Observations#	8	−2.6	10.3	1.4
		June–July–August		
CCM	6	21.4	30.1	24.8
GFDL	6	20.3	28.0	25.0
GISS	4	12.7	21.2	17.3
OSU	8	12.6	16.8	14.5
Observations	8	19.2	28.9	22.3

† Community Climate Model, see Washington & Meehl (1984).
‡ Geophysical Fluid Dynamics Laboratory, see Wetherald & Manabe (1986).
§ Goddard Institute for Space Studies, see Hansen et al. (1984).
¶ Oregon State University, see Schlesinger & Zhao (1989).
# Jenne (1975).

models have been applied perhaps most frequently to the greenhouse climate problem. This problem is dominated by changes in radiative properties in the atmosphere, which the climate models can reliably represent. Much of the uncertainty in model results arises from secondary changes in the climate system that feed back in such a way as to amplify or diminish the initial climatic response to the radiative tendencies. Foremost among these feedback mechanisms are clouds, which will likely change as climate changes but in ways that remain poorly understood.

The most common experiment performed to test the effects of changing atmospheric composition on climate is an idealized one. Instead of calculating the transient response of climate to gradual additions of greenhouse gases, a simpler experiment is usually carried out. Typically, a climate model is run to statistical thermodynamic equilibrium with either preindustrial or contemporary concentrations of CO_2. Then the CO_2 concentration is doubled and, following an adjustment period, a new equilibrium climate is reached. This approach allows for a simplified treatment of oceans, which might take >100 yr to adjust completely to the new climate, and reduces the length of the simulation to a period of only a few decades or so, depending on the model.

Comparisons of the two equilibrium climates can be used to gauge the eventual effect of increasing concentrations of greenhouse gases. In a gross sense, results from this "standard sensitivity experiment" can be interpolated to estimate the response to different concentrations of CO_2. This is possible because the CO_2 radiative forcing has been found to be roughly proportional to the logarithm of CO_2 concentration, and temperature is likely to be approximately linearly related to the forcing. Furthermore, this same idealized experiment can be used to estimate the temperature response to changes in other trace gases by evaluating their radiative importance relative to CO_2. One estimate of the relative radiative importance of other trace gases, based on current growth rates, is given in the last column of Table 1–1. The table shows that the current rates of increase of other trace gases will together approximately double the climatic effects of the current rate of CO_2 increase. This means, roughly speaking, that if current relative rates of emission continue, an "equivalent CO_2 doubling" will occur at a time CO_2 concentrations are only 50% higher than in 1990.

Results from current versions of climate models indicate that globally averaged surface air temperatures will rise by 1.5 to 4.5 K when the CO_2 concentration is doubled in the atmosphere (or, more precisely, for an "equivalent CO_2 doubling"). Most of the differences in the estimates of climate sensitivity are likely due to the different ways in which the models treat clouds (Cess et al., 1989) and to the differences in the control climate surface temperature predicted by the models (apparently the colder the climate, the more sensitive it is to increases in CO_2). Ramanathan (1977) has determined that colder climates are more sensitive primarily because of a complicated nonlinear response of the model climate involving the cloud distribution, the vertical temperature profile, and the global circulation.

The global average climate sensitivity provides a coarse measure of the response of climate to changes in atmospheric composition. As is clear from even a cursory look at climate model results, however, this global average may not accurately reflect the climate change that is predicted to occur on regional scales. As in the simulation of today's climate, the different models may agree on the global scale but show quite significant differences at the regional scale. As an illustration of this, Fig. 1–4 shows the change in zonal

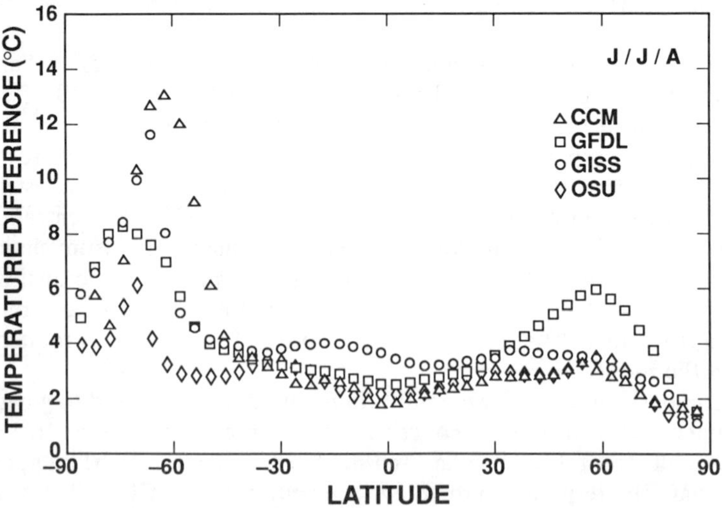

Fig. 1–4. Predicted changes in median zonal surface air temperature in response to a doubling of CO_2, averaged over June, July, and August, vs. latitude for four climate models (adapted from Grotch, 1988). (See Fig. 1–3 for model abbreviation identification.)

surface temperature predicted by four climate models in response to a doubling of CO_2. Similar temperature changes were obtained by Wilson and Mitchell (1987). The models generally agree that the polar regions of the winter hemisphere will experience larger changes than the tropics, but beyond this the detailed responses of the models have little in common.

The previous section of this summary of model results focused on the potential changes in surface air temperatures. Climate, however, consists of a host of other elements, which are also likely to change. Thus, for example, there are expected to be changes in the hydrological cycle, shifts in prevailing wind patterns, and changes in the storm intensity. The analysis of climate model results with regard to these other important meteorological parameters has not progressed beyond very elementary stages. Preliminary studies indicate that there may be even less consistency among model estimates of changes in these other climate parameters than has been found for surface air temperature. From current model projections of precipitation changes, for example, it appears likely that on a global basis, precipitation will generally increase and seasonal patterns of precipitation will change significantly. However, there is no consensus among models as to what specific changes will occur. Evaporation and evapotranspiration should also increase as temperatures rise, but not everywhere since slight shifts in climate zones may counteract the effects of increased temperature. If local increases in evaporation are not offset by sufficiently increased precipitation, decreased soil moisture will result. In some models, increased evaporation amplified the surface warming in some regions, because with dry soils, evaporation can no longer cool the surface. In most regions the models predict a decrease in the portion of winter precipitation that falls as snow. This may significantly affect watersheds that rely on the winter snow pack as a reservoir for water needed later in the year.

Climate models have proven to be powerful tools for studying climate because they permit detailed analysis of feedback processes. The direct effect of increased atmospheric concentration of greenhouse gases is to alter the radiative properties of the atmosphere so as to warm the Earth's surface. But this direct impact gives rise to changes in the climate system that in some cases tend to amplify the initial warming (positive feedbacks) and in other cases to diminish it (negative feedbacks). Analyses of current model results indicate that positive feedbacks dominate, although it is not certain that all feedbacks are being considered. The most important positive feedbacks involve changes in water vapor, cloud altitude and cover, and surface albedo (Hansen et al., 1984; Schlesinger & Mitchell, 1987; Wetherald & Manabe, 1988). It has been estimated, for example, that without feedbacks, the global average temperature would increase by only about 1.2 K following a doubling of atmospheric CO_2. Water vapor feedback approximately doubles this response through the mechanism illustrated in Fig. 1–5. As the climate warms, the absolute humidity increases because the relative humidity changes comparatively little. Because water vapor absorbs infrared radiation, it enhances the effectiveness of the atmospheric greenhouse, which leads to further warming.

The possibility that some of the complex feedbacks in the climate system are not accurately simulated by climate models makes all model projections somewhat uncertain. The most important feedback that may not yet be adequately treated in models involves changes in clouds. Clouds dramatically affect the Earth's radiation balance. Clouds reflect about 20% of the incoming sunlight, which keeps the Earth cooler than it would otherwise be. But clouds also absorb infrared radiation emitted by the surface and the atmosphere itself, enhancing the greenhouse effectiveness of the atmosphere and keeping the Earth warmer than it would otherwise be. Observations (Ramanathan et al., 1989) and theoretical model studies (Cess & Potter, 1987) indicate that the effects of clouds on sunlight absorbtion dominate under current climatic conditions, but the intensity of this effect may shift in an undetermined manner as the climate changes. Most models indicate that as the climate warms, total cloud cover decreases, and the tops of the clouds rise. The change in cloud distribution has the effect of amplifying the initial warming, but these results must be considered preliminary because the parameterizations formulated to account for clouds and their optical properties are highly simplified and therefore may not be accurate.

Improvements in the treatment of clouds and other highly simplified elements of climate models, such as the ocean mixed layer, the surface boundary layer, and the surface hydrology, are needed to reduce the uncertainties that remain in our understanding of climate and the potential equilibrium climate change that will result from imposed changes in atmospheric composition. Furthermore, the current versions of these models cannot yet be used to predict how fast the climate will change. The rate of climate change however, is as important a consideration as the eventual magnitude of the change because society's ability to adapt depends critically on how quickly the changes occur.

The transient response of the climate system to the constantly changing atmospheric composition is primarily determined by how much of the trapped energy is taken up by the oceans. The oceans delay the full climatic response

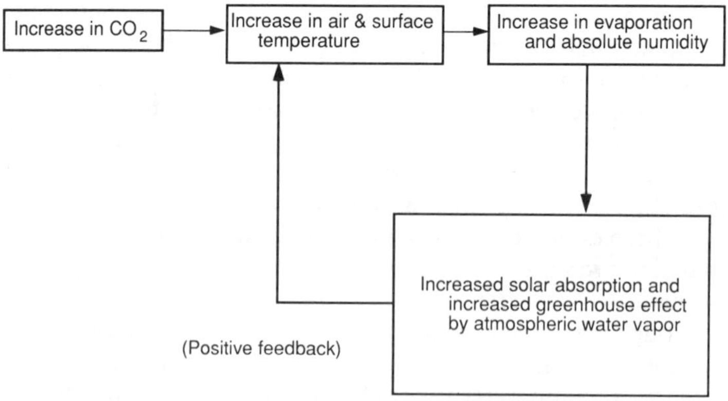

Fig. 1-5. Schematic diagram of the positive humidity feedback, which amplifies climate changes associated with changes in atmospheric composition.

to any forcing by at least a few decades. It is difficult to estimate the effective thermal inertia provided by the oceans because the processes by which heat is mixed and advected by ocean currents from the surface to the deeper layers of the oceans are not quantitatively well understood. Depending on what assumptions are made concerning the ocean circulation, estimates of the response time of the ocean range from a few decades to >100 yr (Hoffert et al., 1980; Harvey & Schneider, 1985; Harvey, 1986; Hansen et al., 1985; Wigley & Schlesinger, 1985).

Even with the uncertainties discussed above, researchers are trying to calculate the time-dependent evolution of future climate. In the most complete simulation yet published, Hansen et al. (1988) suggest that global average temperatures could warm (relatively steadily) by about 0.5 to 3 K by the year 2050, as compared with the climate of the 20th century, depending on whether emissions are stringently controlled or allowed to increase unchecked at rates typical of the 1970s and 1980s. Their most plausible case (although projecting technological changes is inherently uncertain) suggests a warming of about 1.5 K from 1980 to about 2050. Although there is a factor of, perhaps, two in the uncertainty of this estimate, such a change, coupled with the approximately 0.5 K change observed from 1850 to 1980, would move the Earth (and its ecological systems) to sustained global warmth not experienced in the past 100 000 yr. In addition, because the ocean slows the warming rate, even if all emissions were to be curtailed, some further warming is inevitable as the climate equilibrates to the already altered atmospheric composition.

THE CLIMATE RECORD

There is disagreement as to whether the expected CO_2-induced temperature changes are as yet large enough to be distinguished from natural variations in the climate record. Figure 1–6 shows the surface air temperature variations for the Northern Hemisphere for the years 1850 to 1987 as estimated from station data. Although temperatures in the 1980s are estimated to be about 0.5 K higher than they were a century earlier (prior to significant fossil fuel combustion), the trend has by no means been monotonic. In addition to interannual variations of several tenths of a degree, the record shows decadal average temperatures increasing until 1940, then cooling until the 1970s, and then warming again. Climate models predict that temperature should be warming at an increasingly rapid rate (at least initially) in response to continuing emissions of CO_2 and other trace gases. Because of the uncertainties in the response time of the climate system, however, it is not possible to say exactly how much the temperature should already have changed. Thus, the observed temperature trend over the last century cannot now be used to confirm or discredit current estimates of the magnitude of the CO_2 greenhouse effect.

Part of the problem is that the global temperature record is simply not sufficiently accurate to be used for verification of a CO_2-induced signal. There have been changes in measurement methods and problems with "heat

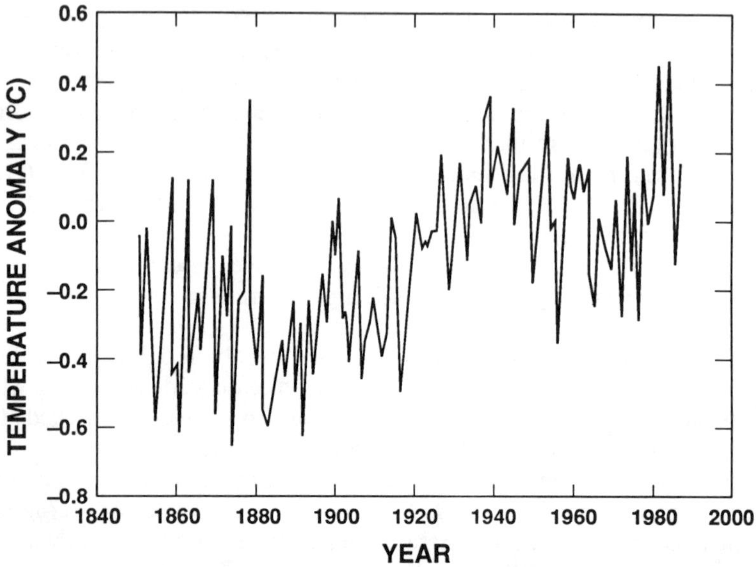

Fig. 1–6. Land-based annual mean surface air temperatures for the Northern Hemisphere. The series is expressed as a departure from the reference period 1951–1970 (Jones, 1989, personal communication; see also Jones, 1988).

island'' effects as cities grow around observation stations, which cause inconsistencies that are difficult to correct (Wigley et al., 1985). There are also large spatial gaps in the data, especially in areal coverage over unpopulated areas and the oceans (Wigley et al., 1985).

The problem of detecting a CO_2-induced climate change is also complicated by other agents that may be affecting climate over the time scales of interest (months to years). For example, episodic volcanic eruptions have loaded the stratosphere with aerosols that may have caused global cooling, in some cases lasting >1 yr. It is possible that solar irradiance has varied enough during the 19th and 20th centuries to have had an impact on climate too. Some of the climatic variation is no doubt due to internal fluctuations within the climate system as the oceans, atmosphere, snow and ice cover, and clouds all interact in a complex and highly nonlinear fashion. All of these and many other factors contribute to the climatic ''noise'' from which we must try to extract the CO_2-induced signal. For this reason, it may be several years before climate change can be specifically and quantitatively attributed to the changes in atmospheric composition.

CONCLUSIONS

Although many questions about how the changing composition of the atmosphere will affect climate remain unanswered, climate models provide strong theoretical reasons to believe that global warming will occur. It seems likely that the warming during the 21st century could be measured in degrees rather than tenths of degrees. However, the details of the regional temperature, precipitation, and soil moisture changes that accompany the warming remain uncertain. Although intensified scientific efforts are being made to develop better estimates of future climatic change, many of the climatological elements of interest to ecologists and farmers are not yet well-simulated and little is known about how they will change in the future.

Changes in precipitation and evaporation will undoubtedly occur, but will vary from region to region. Rising snowlines may decrease the winter snowpack and alter the availability of water for irrigation. Sea level will rise to some extent as the oceans warm and as glaciers melt. This will endanger coastal habitats (initially due to increased sea level during storm surges) and will lead to salt water intrusion into coastal estuaries and aquifers. Early projections were for sea level to rise from about 0.5 to 1.5 m by 2100, which would inundate large areas of low-lying lands; more recent estimates tend toward the lower end of this range or below (Meier, 1990).

We do not know enough to dismiss the possibilities of some geophysical surprises that could either amplify or reduce future climate changes. In such a nonlinear system as climate, it is possible that abrupt shifts from one type of circulation (weather) regime to another could occur quite suddenly. There is in fact some evidence in the paleoclimatic record that ocean currents and atmospheric circulation may have undergone abrupt changes in the past (Broecker, 1987), and it is certainly conceivable that the climate might respond in a highly nonlinear way to changes in atmospheric composition.

The potential impacts of climate change on natural and planned ecosystems will be subject to intense scrutiny in the years to come. Because it is now clear that we can no longer afford to assume that the climate of the past will inevitably be the climate of the future, it would seem appropriate to take into account climate change as an important consideration in planning for our planet's future.

ACKNOWLEDGMENT

We thank S.L. Grotch, who carried out much of the analysis of model results summarized here and helped in the preparation of some of the figures. We also thank P.S. Connell and D.J. Wuebbles for helpful discussions and B.A. Kimball, L.H. Allen, and N.J. Rosenberg for their extensive editorial comments and suggestions.

REFERENCES

Broecker, W.S. 1987. Unpleasant surprises in the greenhouse? Nature (London) 328:123–126.

Cess, R.D., and G.L. Potter. 1987. Exploratory studies of cloud radiative forcing with a general circulation model. Tellus Ser. 39A:460–473.

Cess, R.D., G.L. Potter, J.P. Blanchet, G.J. Boer, S.J. Ghan, J.T. Kiehl, H. Letreut, Z.-X. Li, X.-Z. Liang, J.F.B. Mitchell, J.-J. Morcrette, D.A. Randall, M.R. Riches, E. Roechner, U. Schlese, A. Slingo, K.E. Taylor, W.M. Washington, R.T. Wetherald, and I. Yagai. 1989. Interpretation of cloud-climate feedback as produced by 14 atmospheric general circulation models. Science (Washington, DC) 245:513–516.

Dickinson, R.E., and R.J. Cicerone. 1986. Future global warming from atmospheric trace gases. Nature (London) 319:109–115.

Grotch, S.L. 1988. Regional intercomparison of general circulation model predictions and historical climate data—TR041.DOE/NBB-0084. U.S. Dep. of Energy, Washington, DC. Available from NTIS, Springfield, VA.

Hansen, J., I. Fung, A. Lacis, D. Rind, S. Lebedeff, R. Ruedy, G. Russell, and P. Stone. 1988. Global climate changes as forecast by Goddard Institute for Space Studies three-dimensional model. J. Geophys. Res. 93:9341–9364.

Hansen, J., A. Lacis, D. Rind, G. Russell, P. Stone, I. Fung, R. Ruedy, and J. Lerner. 1984. Climate sensitivity: Analysis of feedback mechanisms. p. 103–163. *In* J.E. Hansen and T. Takahasi (ed.) Climate processes and climate sensitivity (Maurice Ewing Ser. no. 5). Am. Geophysical Union, Washington, DC.

Hansen, J., G. Russell, A. Lacis, I. Fung, D. Rind, and P. Stone. 1985. Climate response times: Dependence on climate sensitivity and ocean mixing. Science (Washington, DC) 229:857–859.

Harvey, L.D.D. 1986. Effect of ocean mixing on the climate response to a CO_2 increase: Analysis of recent model results. J. Geophys. Res. 91:2709–2718.

Harvey, L.D.D., and S.H. Schneider. 1985. Transient climate response to external forcing on 10^0–10^4 year time scales. Part 1: Experiment with globally averaged coupled, atmosphere and ocean energy balance models. J. Geophys. Res. 90:2191–2205.

Hoffert, M.I., A.J. Callegari, and C.-T. Hsieh. 1980. The role of deep sea heat storage in the secular response to climatic forcing. J. Geophys. Res. 85:6667–6679.

Jenne, R.L. 1975. Data sets for meteorological research. NCAR Tech. Note (NCAR-TN/IA-111). Pub. Office of NCAR, Boulder, CO.

Jones, P.D. 1988. Hemispheric surface air temperature variations: Recent trends and an update to 1987. J. Clim. 1:654–660.

MacCracken, M.C., and F.M. Luther (ed.). 1985a. Detecting the climatic effects of increasing carbon dioxide. DOE/ER-0235. U.S. Dep. of Energy, Washington, DC. Available from NTIS, Springfield, VA.

MacCracken, M.C., and F.M. Luther (ed.). 1985b. Projecting the climatic effects of increasing carbon dioxide. DOE/ER-0237. U.S. Dep. of Energy, Washington, DC. Available from NTIS, Springfield, VA.

Meier, M. 1990. Reduced rise in sea level. Nature (London) 343:115–116.

Ramanathan, V. 1977. Interactions between ice-albedo, lapse-rate and cloud-top feedbacks: An analysis of the nonlinear response of a GCM climate model. J. Atmos. Sci. 34:1885–1897.

Ramanathan, V. 1988. The greenhouse theory of climate change: A test by an inadvertent global experiment. Science (Washington, DC) 240:293–299.

Ramanathan, V., L. Callis, R. Cess, J. Hansen, I. Isaksen, W. Kuhn, A. Lacis, F. Luther, J. Mahlman, R. Reck, and M. Schlesinger. 1987. Climate-chemical interactions and effects of changing atmospheric trace gases. Rev. Geophys. 25:1441–1482.

Ramanathan, V., R.D. Cess, E.F. Harrison, P. Minnis, B.R. Barkstrom, E. Ahmad, and D. Hartman. 1989. Cloud-radiative forcing and climate: Results from the earth radiation budget experiment. Science (Washington, DC) 243:57–63.

Ramanathan, V., R.J. Cicerone, H.B. Singh, and J.T. Kiehl. 1985. Trace gas trends and their potential role in climate change. J. Geophys. Res. 90:5547–5566.

Rotty, R.M., and C.D. Masters. 1985. Carbon dioxide from fossil fuel combustion: Trends, resources, and technological implications. p. 63–80. *In* J.R. Trabalka (ed.) Atmospheric carbon dioxide and the global carbon cycle. DOE/ER-0239. U.S. Dep. of Energy, Washington, DC. Available from NTIS, Springfield, VA.

Schlesinger, M.E., and J.F.B. Mitchell. 1987. Climate model simulations of the equilibrium climatic response to increased carbon dioxide. Rev. Geophys. 25:760–798.

Schlesinger, M.E., and Z.-C. Zhao. 1989. Seasonal climatic changes induced by doubled CO_2 as simulated by the OSU atmospheric GCM/mixed-layer ocean model. J. Clim. 2:459–495.

Schneider, S.H. 1989. The greenhouse effect: Science and policy. Science (Washington, DC) 243:771–781.

Trabalka, J.R. (ed.). 1985. Atmospheric carbon dioxide and the global carbon cycle. DOE/ER-0239. U.S. Dep. of Energy, Washington, DC. Available from NTIS, Springfield, VA.

Wang, W.-C., D.J. Wuebbles, W.M. Washington, R.G. Isaacs, and G. Molnar. 1986. Trace gases and other potential perturbations to global climate. Rev. Geophys. 24:110–140.

Washington, W.M., and G.A. Meehl. 1984. Seasonal cycle experiment on the climate sensitivity due to a doubling of CO_2 with an atmospheric general circulation model coupled to a simple mixed-layer ocean model. J. Geophys. Res. 89:9475–9503.

Wetherald, R.T., and S. Manabe. 1986. An investigation of cloud cover change in response to thermal forcing. Clim. Change 8:5–23.

Wetherald, R.T., and S. Manabe. 1988. Cloud feedback processes in a general circulation model. J. Atmos. Sci. 45:1397–1415.

Wigley, T.M.L., J.K. Angell, and P.D. Jones. 1985. Analysis of the temperature record. p. 55–90. *In* M.C. MacCracken and F.M. Luther (ed.) Detecting the climatic effects of increasing carbon dioxide. DOE/ER-0235. U.S. Dep. of Energy, Washington, DC. Available from NTIS, Springfield, VA.

Wigley, T.M.L., and M.E. Schlesinger. 1985. Analytic solution for the effect of increasing CO_2 on global mean temperature. Nature (London) 315:649–652.

Wilson, C.A., and J.F.B. Mitchell. 1987. A doubled CO_2 climate sensitivity experiment with a global climate model including a simple ocean. J. Geophy. Res. 92:13 315–13 343.

Wuebbles, D.J., and J. Edmonds. 1988. A primer on greenhouse gases. DOE/NBB0083. U.S. Dep. of Energy, Washington, DC. Available from NTIS, Springfield, VA.

2 The Carbon Dioxide/Trace Gas Greenhouse Effect: Greatly Overestimated?[1]

Sherwood B. Idso

U.S. Water Conservation Laboratory
Phoenix, Arizona

ABSTRACT

Most general circulation models of the atmosphere predict that a 300 to 600 μmol CO_2/mol air doubling of the atmospheric CO_2 concentration will increase Earth's mean surface air temperature by about 4 °C, and that concomitant increases in certain other trace gases will produce another 4 °C warming. Because of a number of demonstrable model deficiencies, however, this conclusion should not be accepted as definitive. For one thing, many well-known processes of importance to world climate are not properly represented in the models. A number of other pertinent phenomena are not even included; and the models fail to temper their calculations in accordance with the requirements of certain large-scale empirical constraints. As a result, there is reason to believe that current climate model predictions of the strength of the CO_2/tracc gas greenhouse effect may be an order of magnitude too large.

It is certainly no secret that I have a significantly different opinion of the CO_2/trace gas greenhouse effect than the one described by Taylor and MacCracken (Chapter 1, this book). Consequently, it will be my purpose in this chapter to introduce the rationale for a considerably more optimistic view of the steadily rising CO_2 content of Earth's atmosphere than that generally portrayed in the scientific literature and by the public media.

CLIMATE MODEL SHORTCOMINGS

First of all, I must say that my primary problem with the general circulation models of the atmosphere, or GCMs, is not what they do, but what they *don't* do. In fact, with respect to Schlesinger and Mitchell's (1987) state-of-the-art assessment of the climate modeling enterprise, I have stated in a

[1] Contribution from the USDA-ARS.

letter to the editor of *Climatic Change* (Idso, 1988b) that "I concur in every single word of their conclusions and suggested goals of future research." Nevertheless, I find it difficult to accept their current predictions.

My first concern is one of simple logic, as articulated by Reck (1982) at the 1981 LaJolla Institute workshop: The Responsible Interpretation of Atmospheric Models and Related Data. At that meeting, Reck made the important point that "confidence in the predictions of atmospheric models is limited, (i) by the uncertainty in our knowledge of the existing atmospheric system as it is defined by input data, and (ii) by the ability of the model to describe all pertinent atmospheric processes." She then went on to indicate that to overcome these limitations, "it is necessary that the model sensitivity to parameter perturbation be equivalent to that observed in the real world" and that "this requires that the model include all necessary physical mechanisms as well as pertinent coupling schemes between parameters." And I just don't believe that any climate model in use today even comes close to meeting these requirements.

By way of explanation, the climate models in use at the time of the LaJolla Institute workshop, though complex for their day, were extremely crude with respect to reality. As Hansen et al. (1981) noted in a seminal paper published that same year, climate model predictions of the late 1970s and early 1980s were based on "rudimentary abilities to model a complex system with many non-linear processes"; whereas Kellogg (1980) went so far as to state that "no climate model, however sophisticated, can hope to include all the factors and interactions of the planetary system."

As is characteristic of the times in which we live, however, the intervening years have seen much progress in these areas. Nevertheless, in their state-of-the-art report published in May of 1987, Schlesinger and Mitchell concluded that there was still "an inherent limitation in our ability to validate the accuracy of GCM perturbation simulations, which thereby affects our confidence in the accuracy of the GCM simulations of CO_2-induced climate change."

In expanding on this statement, Schlesinger and Mitchell noted that contemporary climate models "frequently employ treatments of dubious merit, including prescribing the oceanic heat flux, ignoring the oceanic heat flux, and using incorrect values for the solar constant." They point out that "such approximations indicate that the models are physically incomplete and/or have errors in the included physics." Schlesinger and Mitchell note further that "CO_2-induced climate changes simulated by different GCMs show many quantitative and even qualitative differences," and conclude by saying that "we know that not all of these simulations can be correct, and perhaps all could be wrong."

In an even newer intercomparison of climate model predictions, Grotch (1988) found that the climate models did not always agree well with each other, and in some instances they failed to adequately reproduce the proper annual cycle of climate over the world. Such was also the take-home message of the model intercomparison of Gutowski et al. (1988), where certain regional control-climte energy fluxes actually changed in different directions after doubling the atmospheric CO_2 content. And in updating the study of Schlesinger and Mitchell, Cess and Potter (1988) indicated that "more recent GCM simulations have actually broadened [the] range of apparent disagreement." Hence, as stated by Ramanathan (1988), our "theoretical understanding of the [Earth's] climate system is by no means complete." And as suggested by Kellogg (1980) almost a decade earlier, perhaps it never will be.

Although I am not nearly as pessimistic as Kellogg in this regard, in that I believe that the climate modeling enterprise will eventually succeed, I'm convinced it will be a long time in coming; the Earth's climate system is profusely rich in interacting components, about many of which we still know relatively little.

Consider, for instance, the statement of Schlesinger and Mitchell that present day GCMs "do not resolve all of the physical processes that may be of importance to climate and climate change and which span the 14 orders of magnitude from the planetary scale (10^7 m) to the cloud microphysical scale (10^{-6} m)." Noting that "contemporary computers permit the resolution of physical processes over only 2 orders of magnitude," they point out that "even a thousandfold increase in computer speed, which is not projected to occur within this century, would allow the resolution of only 1 more order of magnitude!"

CLIMATE EFFECTS OF CLOUDS

As just one example of how important some of these unresolved physical processes may be, consider the effect of increasing air temperature on cloud liquid water content (Paltridge, 1980; Charlock, 1981, 1982), which is still missing from most GCMs. Because a warmer troposphere would contain more water vapor and, hence, more condensate, the clouds in a warmer world would contain more liquid water than those of the present. And because the infrared opacities of low- and midlevel clouds saturate very quickly (Betts & Harshvardhan, 1987), such an increase in cloud liquid water content would produce a negative or stabilizing effect on the initial CO_2-induced warming. That is, because the heat-conserving "greenhouse" properties of these clouds are already near their maximum, while their reflectances for solar radiation still have the capacity to considerably increase (Somerville & Remer, 1984), the net effect of likely changes in both parameters due to an increase in cloud liquid water content is to produce a significant impetus for cooling. In fact, when observational estimates (Feigelson, 1978) of the temperature dependence of cloud liquid water content were first employed in a radiative-convective climate model by Somerville and Remer (1984), they found that "the surface temperature change caused by doubling CO_2 [was] reduced by about one half," which has subsequently been confirmed by a more complex GCM as well (Roeckner et al., 1987).

In the case of optically thin high cirrus clouds, however, just the opposite effect occurs, with more outgoing thermal radiation being trapped within the atmosphere than there is incoming solar radiation being reflected. As a result, Roeckner (1988) has recently noted that the ultimate sign of the total planetary response to an increase in cloud liquid water content is "crucially dependent on the spatial distribution of cloud amount and cloud optical properties," which has prompted Schlesinger (1988) to conclude that "because of the complexity of cloud processes in nature, many additional modelling and observational studies will be required to determine the sign of the actual cloud optical depth feedback, let alone its magnitude." And as a result of this and a number of other associated problems, Ramanathan (1988) has wisely stated that "since clouds are treated primitively in GCMs, it is premature to make reliable inferences from GCM studies."

BIOLOGICAL FEEDBACK MECHANISMS

Clearly, then, any number of physical phenomena, many of which, I am sure, have yet to even be discovered, could readily alter GCM predictions of climatic change. However, there is an even more compelling reason to question their predictions: they are essentially devoid of any *biological* input (Lovelock, 1988). And as the scientific community is just now coming to fully realize (Kerr, 1988; Lindley, 1988; Pearce, 1988), it is impossible to adequately understand the temperature history of the Earth without a good dose of what Lovelock (1979) has called *geophysiology*, or the study of the planet as a complex but integrated living entity, actively controlling the physical and chemical states of the environment in such a way as to provide for its own continued existence, which in very simple terms is the basic tenet of what has come to be known as the *Gaia Hypothesis*.

On a very broad scale, the Gaian view of the world attributes the relative stability of Earth's surface air temperature over the eons, i.e., the faint early sun paradox (Sagan & Mullen, 1972; Owen et al., 1979; Lovelock & Whitfield, 1982), to the important role of the planet's biota in steadily removing CO_2 from the atmosphere as the energy output of the sun has gradually risen (by perhaps 30%) over Earth's 4.5 billion year history (Lovelock, 1988). And as I have shown elsewhere (Idso, 1988d), this phenomenon can only operate as postulated, i.e., to keep the Earth's mean surface air temperature relatively stable by progressively decreasing the atmosphere's total greenhouse effect via the biological removal of CO_2 from the air, if the climatic sensitivity of the Earth to changes in the CO_2 content of the atmosphere is fully an order of magnitude *less* than that suggested by current state-of-the-art GCMs.

But how could the models possibly be so wrong? Consider, for a moment, the abundant phytoplankton of the world's oceans. As they photosynthesize, they produce a substance called dimethylsulfoniopropionate (Vairavamurthy et al., 1985), which disperses throughout the surface waters of the oceans when they either die (Nguyen et al., 1988) or are eaten by zooplankton (Dacey & Wakeham, 1986). Then, a portion of it slowly decomposes to produce dimethyl sulfide or DMS (Turner et al., 1988), which is the most abundant volatile S compound in seawater (Wakeham et al., 1984).

A portion of the DMS thus introduced to the surface waters of the world's oceans diffuses into the atmosphere, where it is rapidly oxidized by ubiquitous hydroxyl radicals and converted into sulfuric and methanesulfonic acid particles (Andreae, 1986), which are carried aloft to ultimately function as cloud condensation nuclei (Bates et al., 1987), some of which produce new clouds and some of which dramatically increase the albedos of preexistent clouds (Coakley et al., 1987; Charlson et al., 1987). This phenomenon thus produces a cooling effect which, in the words of Lovelock (1988), "is comparable in magnitude with that of the carbon dioxide greenhouse, but in opposition to it." Together, these two mechanisms may thus constitute a global "Gaian thermostat," which effectively works to stabilize the temperature of the planet and prohibit the radical greenhouse warming predicted by current GCMs (Idso, 1989b).

REAL-WORLD CLIMATIC CONSTRAINTS

Clearly, however, the whole story is considerably more complex than what I have been able to touch on in this brief chapter. It additionally includes, for example, the empirically observed responsiveness of various Earth surfaces to radiative forcing (Idso, 1984b, 1989b; Idso et al., 1986), various empirically observed limits to surface air temperature rise (Idso, 1982, 1985, 1989a), the radiative properties of competing atmospheric greenhouse constituents (Idso, 1983, 1984a), the means by which the temperature history of the Earth is inferred from actual temperature measurements (Jones et al., 1986a, b, c; Hansen & Lebedeff, 1987; Wood, 1988; Balling & Idso, 1989b), the mechanics of Quaternary glaciations (Saigne & Legrand, 1987; Legrand et al., 1988; Idso, 1988c; Roger & Wilson, 1989), the nature of the most recent "Little Ice Age" (Grove, 1988; Idso, 1988a), the empirically observed surface temperatures and partial pressures of CO_2 on Mars and Venus (Idso, 1988d), plus a number of other matters related to real-world observations (Idso, 1986a, b, 1987a, b).

CONCLUSIONS

All of these many facets of the CO_2/trace gas greenhouse effect problem are important and must be dealt with in depth, if the issue is ever to be properly resolved. For now, it is merely hoped that this brief glimpse of the complex problem has been sufficient to call into question the primary prediction of current state-of-the-art climate models. Being as approximate and incomplete as even their own creators acknowledge them to be, no one should be asked to believe that contemporary model predictions are an accurate representation of real-world climate dynamics. To so affirm at this stage of our understanding, in my opinion, is to execute a leap of faith that is totally unjustified.

REFERENCES

Andreae, M.O. 1986. The ocean as a source of atmospheric sulphur compounds. p. 331–362. *In* P. Buat-Menard (ed.) The role of air-sea exchange in geochemical cycling. Reidel, Dordrecht, Netherlands.

Balling, R.C., Jr., and S.B. Idso. 1989. Historical temperature trends in the United States and the effect of urban population growth. J. Geophys. Res. 94:3359–3363.

Bates, T.S., R.J. Charlson, and R.H. Gammon. 1987. Evidence for the climatic role of marine biogenic sulphur. Nature (London) 329:319–321.

Betts, A.K., and Harshvardhan. 1987. Thermodynamic constraint on the cloud liquid water feedback in climate models. J. Geophys. Res. 92:8483–8485.

Cess, R.D., and G.L. Potter. 1988. A methodology for understanding and intercomparing atmospheric climate feedback processes in general circulation models. J. Geophys. Res. 93:8305–8314.

Charlock, T.P. 1981. Cloud optics as a possible stabilizing factor in climate change. J. Atmos. Sci. 38:661–663.

Charlock, T.P. 1982. Cloud optical feedback and climate stability in a radiative-convective model. Tellus 34:245–254.

Charlson, R.J., J.E. Lovelock, M.O. Andreae, and S.G. Warren. 1987. Oceanic phytoplankton, atmospheric sulfur, cloud albedo and climate. Nature (London) 326:655–661.

Coakley, J.A., R.L. Berstein, and P.A. Durkee. 1987. Effect of shipstack effluent on cloud reflectivity. Science (Washington, DC) 237:1020–1022.

Dacey, J.W.H., and S.G. Wakeham. 1986. Oceanic dimethylsulfide: Production during zooplankton grazing on phytoplankton. Science (Washington, DC) 233:1314–1316.

Feigelson, E.N. 1978. Preliminary radiation model of a cloudy atmosphere: 1. Structure of clouds and solar radiation. Contrib. Atmos. Phys. 51:203–229.

Grotch, S.L. 1988. Regional intercomparisons of general circulation model predictions and historical climate data—TR041. DOE/NBB-0084. U.S. Dep. of Energy, Washington, DC. Available from NTIS, Springfield, VA.

Grove, J.H. 1988. The little ice age. Methuen, London, UK.

Gutowski, W.J., D.S. Gutzler, D. Portman, and W.-C. Wang. 1988. Surface energy balance of three general circulation models: Current climate and response to increasing atmospheric CO_2—TR042. U.S. Dep. of Energy, Washington, DC. Available from NTIS, Springfield, VA.

Hansen, J., D. Johnson, A. Lacis, S. Lebedeff, P. Lee, D. Rind, and G. Russell. 1981. Climate impact of increasing atmospheric carbon dioxide. Science (Washington, DC) 213:957–966.

Hansen, J., and S. Lebedeff. 1987. Global trends of measured surface air temperature. J. Geophys. Res. 25:13 345–13 372.

Idso, S.B. 1982. Temperature limitation by evaporation in hot climates and the greenhouse effects of water vapor and carbon dioxide. Agric. Meteorol. 27:105–109.

Idso, S.B. 1983. CO_2 as an *inverse* greenhouse gas. p. 232–239. *In* I.J. Spiro and R.A. Mollicone (ed.) Infrared technology IX. Proc. Society of Photo-Optical Instrumentation Engineers, San Diego, CA. 23–25 August. SPIE, Bellingham, WA.

Idso, S.B. 1984a. What if increases in atmospheric CO_2 have an *inverse* greenhouse effect? I. Energy balance considerations related to surface albedo. J. Climatol. 4:399–409.

Idso, S.B. 1984b. An empirical evaluation of earth's surface air temperature response to radiative forcing, including feedback, as applied to the CO_2-climate problem. Archiv. Meteorol. Geophys. Bioclim. Ser. B 34:1–19.

Idso, S.B. 1985. An upper limit to global surface air temperature. Archiv. Meteorol. Geophys. Bioclim. Ser. A 34:141–144.

Idso, S.B. 1986a. My response to the concluding 1984 issue of *Climatic Change* dealing with the CO_2/climate controversy. CO_2/Climate Dialogue 1(1):7–66.

Idso, S.B. 1986b. My response to Appendix B—"Review of the recent carbon dioxide-climate controversy"—of the DOE-sponsored state-of-the-art volume: Projecting the climatic effects of increasing carbon dioxide. CO_2/Climate Dialogue 1(2):29–37.

Idso, S.B. 1987a. A clarification of my position on the CO_2-climate connection. Clim. Change 10:81–86.

Idso, S.B. 1987b. Carbon dioxide and climate: The legacy of logic. CO_2/Climate Dialogue 2(1):9–17.

Idso, S.B. 1988a. Greenhouse warming or Little Ice Age demise: A critical problem for climatology. Theoret. Appl. Climatol. 39:54–56.

Idso, S.B. 1988b. Me and the modelers: Perhaps not so different after all. Clim. Change 12:93.

Idso, S.B. 1988c. Carbon dioxide and climate in the Vostok ice core. Atmos. Environ. 22:2341–2342.

Idso, S.B. 1988d. The CO_2 greenhouse effect on Mars, Earth, and Venus. Sci. Total Environ. 77:291–294.

Idso, S.B. 1989a. An upper limit to the greenhouse effect of Earth's atmosphere. Theoret. Appl. Climatol. 40:171–174.

Idso, S.B. 1989b. On the stability of earth's climate. Theoret. Appl. Climatol. 3:177–178.

Idso, S.B., K.L. Clawson, and M.G. Anderson. 1986. Foliage temperature: Effects of environmental factors, with implications for plant water stress assessment and the CO_2/climate connection. Water Resour. Res. 22:1702–1716.

Jones, P.D., S.C.B. Raper, R.S. Bradley, H.F. Diaz, P.M. Kelly, and T.M.L. Wigley. 1986a. Northern hemispheric surface air temperature variations: 1851–1984. J. Clim. Appl. Meteorol. 25:161–179.

Jones, P.D., S.C.B. Raper, and T.M.L. Wigley. 1986b. Southern hemispheric surface air temperature variations: 1851–1984. J. Clim. Appl. Meteorol. 25:1213–1230.

Jones, P.D., T.M.L. Wigley, and P.B. Wright. 1986c. Global temperature variations between 1861 and 1984. Nature (London) 322:430–434.

Kellogg, W.W. 1980. Modeling future climate. Ambio 9:216–221.

Kerr, R.A. 1988. No longer willful, Gaia becomes respectable. Science (Washington, DC) 240:393–395.

Legrand, M.R., R.J. Delmas, and R.J. Charlson. 1988. Climate forcing implications from Vostok ice-core sulphate data. Nature (London) 344:418–420.

Lindley, D. 1988. Is the earth alive or dead? Nature (London) 332:483–484.

Lovelock, J.E. 1979. Gaia: A new look at life on earth. Oxford Univ. Press, Oxford, UK.

Lovelock, J.E. 1988. The ages of Gaia: A biography of our living earth. W.W. Norton & Co., New York.

Lovelock, J.E., and M. Whitfield. 1982. Life span of the biosphere. Nature (London) 296:561–563.

Nguyen, B.C., S. Belviso, N. Mihalopoulos, J. Gostan, and P. Nival. 1988. Dimethyl sulfide production during natural phytoplanktonic blooms. Marine Chem. 24:133–141.

Owen, T., R.D. Cess, and V. Ramanathan. 1979. Enhanced CO_2 greenhouse to compensate for reduced solar luminosity on early Earth. Nature (London) 277:640–642.

Paltridge, G.W. 1980. Cloud-radiation feedback to climate. Q. J. R. Meteorol. Soc. 106:895–899.

Pearce,F. 1988.Gaia: A revolution comes of age. New Sci. 117(1604):32–33.

Ramanathan, V. 1988. The greenhouse theory of climate change: A test by an inadvertent global experiment. Science (Washington, DC) 240:293–299.

Reck, R.A. 1982. Confidence in climate models including those with suspended particles. p. 23–33. *In* R.A. Reck andJ.R. Hummel (ed.) AIP Conf. Proc. No. 32: Interpretation of climate and photochemical models, ozone and temperature measurements. American Institute of Physics, New York, NY.

Roeckner, E. 1988. Negative or positive cloud optical depth feedback? Nature (London) 335:304.

Roeckner, E., U. Schlese, J. Biercamp, and P. Loewe. 1987. Cloud optical depth feedbacks and climate modeling. Nature (London) 329:138–140.

Roger, T., and S. Wilson. 1989. Carbon dioxide and climate in the Vostok ice core: Why does the system oscillate? Atmos. Environ. 22:2637–2638.

Sagan, C., and G. Mullen. 1972. Earth and Mars: Evolution of atmospheres and surface temperatures. Science (Washington, DC) 177:52–56.

Saigne, C., and M. Legrand. 1987. Measurements of methanesulphonic acid in Antarctic ice. Nature (London) 330:240–242.

Schlesinger, M.E. 1988. Negative or positive cloud optical depth feedback? Nature (London) 335:303–304.

Schlesinger, M.E., and J.F.B. Mitchell. 1987. Climate model simulations of the equilibrium climatic response to increased carbon dioxide. Rev. Geophys. 25:760–798.

Somerville, R.C.J., and L.A. Remer. 1984. Cloud optical thickness feedbacks in the CO_2 climate problem. J. Geophys. Res. 89:9668–9672.

Turner, S.M., G. Malin, P.S. Liss, D.S. Harbour, and P.M. Holligan. 1988. The seasonal variation of dimethyl sulfide and dimethylsulfoniopropionate concentrations in nearshore waters. Limnol. Oceanogr. 33:364–375.

Vairavamurthy, A., M.O. Andreae, and R.L. Iverson. 1985. Biosynthesis of dimethylsulfide and dimethylpropiothetin by *Hymenomonas carterae* in relation to sulfur source and salinity variations. Limnol. Oceanogr. 30:59–70.

Wakeham, S.G., B.L. Howes, and J.W.H. Dacey. 1984. Dimethyl sulfide in a stratified coastal salt pond. Nature (London) 310:770–772.

Wood, F.B. 1988. Comment: On the need for validation of the Jones et al. temperature trends with respect to urban warming. Clim. Change 12:297–312.

3 Greenhouse Gas Emissions Related to Agriculture and Land-Use Practices[1]

Lauretta M. Burke

The Bruce Company
Washington, DC

Daniel A. Lashof

U.S. Environmental Protection Agency
Washington, DC

ABSTRACT

The effects of increasing trace gas concentrations and concomitant climate change on agriculture are likely to be substantial. With cropland and pasture now covering $<30\%$ of the Earth's land surface, agricultural activity is also a significant factor in producing the observed increases in the greenhouse gases CO_2, CH_4, and N_2O. Land clearing for agriculture and other purposes is responsible for 10 to 30% of total net CO_2 emissions; the rest is due to fossil fuel combustion. In addition, intentional burning of agricultural wastes, grasslands, and forests makes a significant contribution to global emissions of CO, CH_4, NO_x, and N_2O. Methane emissions from anaerobic respiration in rice (*Oryza sativa* L.) paddies and domestic animal rumens account for 30 to 50% of the global total, making agriculture the dominant anthropogenic source of this gas. The amount of N_2O emitted as a result of N fertilizer applications is highly uncertain, but may be on the order of 10% of total N_2O emissions. Future agricultural greenhouse gas emissions will be affected by population growth, economic development, and agricultural practices. Greenhouse gas emissions are likely to increase substantially in the future unless steps are taken to control them. Investigating potential approaches to reducing these emissions while expanding production presents a major challenge to the agricultural research community.

Agriculture and climate are inextricably related. The dependence of agriculture on climate has been appreciated for millenia, whereas the possible contribution of agriculture to altering climate is just beginning to be recognized.

[1] This research was supported by the U.S. Environmental Protection Agency, 401 M Street SW, Washington, DC 20460.

Agriculture and related activities, including clearing forests, burning plant matter, cultivating rice (*Oryza sativa* L.), raising livestock, and using nitrogenous fertilizer, contribute greenhouse gases to the atmosphere.

This chapter will cover the agricultural activities that contribute to the emission of greenhouse gases; the trends that are likely to affect future emissions; and the uncertainties surrounding emissions estimates. Research needs are also outlined for improving emissions estimates and for developing methods to reduce the emissions of greenhouse gases from agriculture.

REPORTS TO CONGRESS

Concern about increasing concentrations of greenhouse gases in the atmosphere and the resulting potential for global warming prompted the U.S. Congress to request two reports from the U.S. Environmental Protection Agency (USEPA) on the subject of climate change. One report, *The Potential Effects of Global Climate Change on the United States* (Smith & Tirpak, 1989), examines the effect of climate change on resources in the USA, including agriculture, forests, and unmanaged ecosystems. The other report, *Policy Options for Stabilizing Global Climate* (Lashof & Tirpak, 1990), examines policy options that, if implemented, would limit the buildup of greenhouse gases in the atmosphere. The "Stabilizing Report" includes an examination of the potential for and effect of reducing the emission of greenhouse gases from agricultural sources and land-use change. This chapter is based largely on research conducted for that report.

THE GREENHOUSE GASES

Greenhouse gases, which occur naturally in the atmosphere, warm the earth by absorbing infrared radiation emitted from the surface (see Taylor & MacCracken, Chapter 1, this book). Agricultural activities emit a wide range of greenhouse gases to the atmosphere and thus contribute to the increasing concentrations of these gases. The clearing of forested land results in the emission of CO_2 from decreased C storage in biomass (nonfossil organic matter) and N_2O from exposed soils. Burning of biomass results in the emission of CH_4, N_2O, NO_x and CO, in addition to CO_2. Livestock and rice cultivation contribute CH_4 to the atmosphere and the use of nitrogenous fertilizer results in increased emission of N_2O from soils.

Due to differing physical properties and atmospheric lifetimes, the various greenhouse gases have different greenhouse warming potentials. The contribution of an individual gas to increases in the greenhouse effect depends on both its change in atmospheric concentration and its radiative forcing per additional molecule in the atmosphere. *Radiative forcing* is defined as the change in the downward infrared flux measured at the tropopause in the absence of feedbacks. At the 1980s levels, increases in CO_2 concentra-

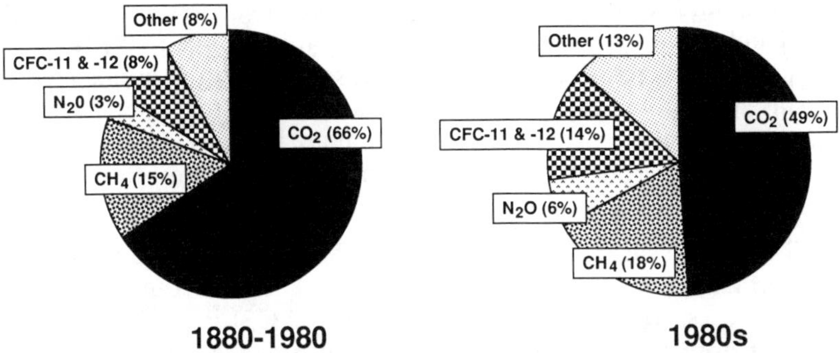

Fig. 3-1. Relative contributions to the increase in the greenhouse effect. Greenhouse gases other than CO_2 account for about half of the increases in the greenhouse effect in the 1980s. The "other" category includes halons, tropospheric O_3, and stratospheric water vapor (sources—1880-1980: Ramanathan et al., 1985; 1980s: Hansen et al., 1988).

tions account for about half of recent increases in the greenhouse effect. Figure 3-1 illustrates the relative contribution of the major greenhouse gases.

Because of the overlap between agricultural activities and greenhouse gases (i.e., individual activities contribute several gases, and concentration increases of each gas are the product of many activities), there is no obvious, straight-forward organization for presentation of this subject. We have chosen to discuss emissions related to land-clearing and biomass burning first; we then discuss the other sources of CH_4 and N_2O.

LAND-USE CHANGE

Land-clearing is probably the agriculture-related activity that has made the largest contribution to the greenhouse gas buildup. The history of atmospheric CO_2, as evidenced by ice core records (Fig. 3-2), shows that CO_2

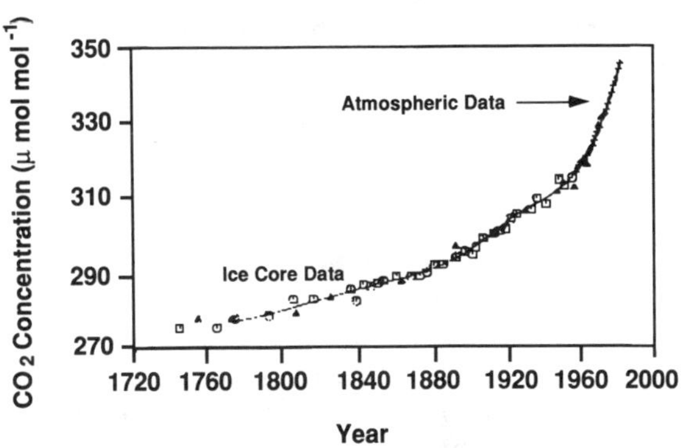

Fig. 3-2. Observed CO_2 concentration. The history of atmospheric CO_2 presented here is based on ice core measurements (squares and triangles) and atmospheric measurements (crosses). The data show that CO_2 began to increase in the 1800s with the conversion of forest to agricultural land. The rapid rise since the 1950s, due primarily to fossil fuel combustion, is at a rate unprecedented in the ice core record (sources—Neftel et al., 1985; Friedli et al., 1986; Keeling, personal communication; as cited in Siegenthaler & Oeschger, 1987).

began to increase in the 1800s, when fossil fuel use was minimal. This increase is attributable to the conversion of forest to agricultural land. The rapid rise since the 1950s is primarily due to fossil fuel combustion (Houghton et al., 1983).

Carbon Dioxide Emissions from Deforestation

When land is cleared, CO_2 is released through the burning or decomposition of plant matter and the oxidation of organic matter in the soil. Conversion of forest to cropland results in decreased C storage, because forest ecosystems store 20 to 100 times more C per unit area than cropland (Houghton et al., 1983).

Global forest area has decreased by 15% since 1850, primarily due to conversion to cropland (Houghton et al., 1983). It is estimated that about 11.3 million ha of tropical forest are lost each year, whereas only 1.1 million ha are reforested (FAO, 1986). As a result there is a net flux of CO_2 from the terrestrial biosphere to the atmosphere, estimated at 0.4 to 2.6 Pg in 1980 (Houghton et al., 1987; Detwiler & Hall, 1988).[2] The wide range of the estimates reflects uncertainties regarding the actual rate of land clearing and the amount of standing biomass on the land before clearing. Research is underway to narrow this uncertainty range through remote sensing (Tucker et al. 1986; Woodwell et al., 1986) and through atmospheric modeling to constrain the distribution of CO_2 sources and sinks (Fung, 1986; Keeling & Heimann, 1986; Tans et al., 1989). Recent remote sensing results indicate that land clearing rates may have incresed dramatically during the 1980s, suggesting that current emissions are more likely to be near the higher end of the range given above (A.W. Setzer, 1989, personal communication; Meyers, 1989). The results of atmospheric models, however, indicate that such a large flux may be inconsistent with the observed pattern of CO_2 concentrations, which may suggest that the lower end of the range is more probable.

With the current uncertainties the net release of C from deforestation accounts for approximately 10 to 30% of annual anthropogenic CO_2 emissions to the atmosphere. Of the estimated source, half was produced by only five countries in 1980: Brazil, Indonesia, Columbia, the Ivory Coast, and Thailand (Houghton et al., 1987).

Biomass Burning

The burning of biomass generates substantial trace gas emissions due to the uncontrolled, incomplete nature of the combustion, and due to the relatively high fuel N content. Globally, biomass burning contributes approximately 10 to 20% of total annual CH_4 emission, 20 to 40% of CO emissions, 5 to 15% of N_2O emissions, and 10 to 35% of NO_x emissions, in addition to contributing to atmospheric CO_2 emissions (Crutzen et al., 1979; WMO, 1985; Logan, 1983; Stevens & Engelkemeir, 1988; Andreae et al.,

[2] 1 petagram = 1 Pg = 10^{15} g = 1 billion Mg = 1 gigaton.

1988). Agricultural burning due to shifting agriculture and the burning of agricultural wastes is estimated to account for $>50\%$ of all biomass burned annually (Crutzen et al., 1979). Uncertainties in estimates of trace gas emissions from biomass burning result from limited data on types and amounts of biomass burned and highly variable emissions per unit of biomass burned.

Nitrous Oxide Emissions from Land Disturbance

Rapid deforestation in the tropics may also be a significant source of increased N_2O emissions from soils. Recent studies have shown enhanced N_2O fluxes to the atmosphere from cleared areas in temperate forest and elevated N_2O concentrations in groundwater adjacent to cut watersheds, suggesting that additional N_2O is being transported off site and emitted where there is no detector (Bowden & Borman, 1986). Clearing by burning has also been shown to lead to elevated N_2O emissions from soils for an extended period after the burn (Anderson et al., 1988; P. Matson, 1988, personal communication). As yet, no estimate of the contribution of deforestation to the global N_2O budget has been attempted.

Trends

Growing populations and economic growth have put increasing pressure on limited forest and land resources, particularly in the developing world. Continued reliance on unsustainable practices such as large-scale conversion of forest to pasture, shifting agriculture with ever-shortening fallow periods, and over-harvest of forest resources will result in increasing emissions of greenhouse gases in addition to depletion of the resource base. In the absence of initiatives to alter this trend, these practices are likely to decline only through exhaustion of the available forests. In the meantime, about 100 Pg C could be emitted to the atmosphere during the next century (about 20 yr worth of fossil fuel emissions at current levels).

The analysis performed for USEPA's report on *Policy Options for Stabilizing Global Climate* included three scenarios for future deforestation/reforestation rates. These scenarios are intended to provide a range of plausible future emissions paths consistent with a range of economic and policy assumptions adopted for other sectors. They were not designed to bound the full range of possibilities, nor should the scenario that falls in the middle be considered a best guess. One scenario (the *slowly changing world,* or SCW) assumes low rates of global economic growth, and as a result, rapid population growth and unsustainable agricultural practices continue. For this scenario regional deforestation rates were assumed to increase with the rate of population growth in Asia, Latin America, and Africa, respectively. Global clearing increases from 11 million ha yr^{-1} in 1980 to 34 million ha yr^{-1} in 2047, at which point all the unprotected forests in Asia are exhausted. Under this scenario, annual C emissions rise from 0.7 to 2 Pg C by 2047 (Fig. 3–3, Lashof & Tirpak, 1990).

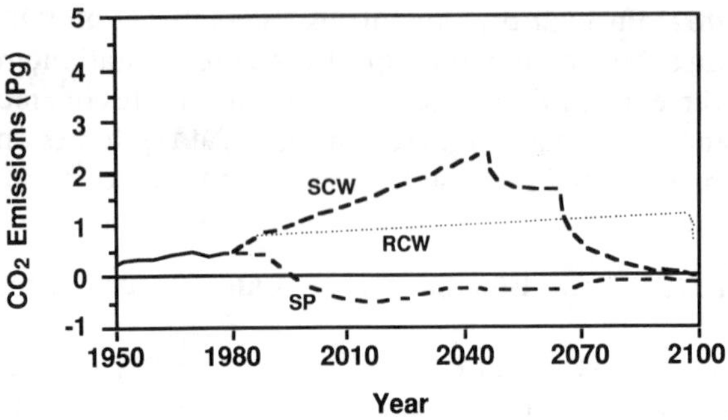

Fig. 3-3. Scenarios of global annual CO_2 emissions from deforestation. (i) Deforestation is most rapid under the slowly changing world (SCW) scenario. The SCW assumes a low rate of economic growth, and as a result, population growth and unsustainable agricultural practices continue. Under this scenario, the deforestation rate is tied to population growth. Carbon dioxide emissions decline when all of the forests in Asia, followed by Latin America and Africa, are exhausted. (ii) Under the rapidly changing world (RCW) scenario, increased economic growth and technological change reduces the pressure on forests compared with the SCW and results in a lower rate of deforestation. Total emissions are similar under the SCW and RCW scenarios, but are spread out over more time in the RCW. (iii) A stabilizing policy (SP) scenario was analyzed, which assumes that deforestation stops by 2025 and that >1000 million ha of land are reforested by 2100. Under this scenario, forests become a net sink for C by 2000 (source—Lashof & Tirpak, 1989).

A second scenario assumes more rapid economic growth and technological change (the *rapidly changing world,* or RCW). Under the RCW scenario, improved agricultural practices and substitution of modern fuels for fuelwood ease the pressure on forests compared with the SCW. Nonetheless, deforestation continues, primarily as a result of commercial logging, ranching, and land speculation. Deforestation is assumed to increase at current rates of growth, reaching 15 million ha yr^{-1} by 2097, when the unprotected forests of Latin America are exhausted. Total C emissions are similar to the SCW case, but are spread out over a longer period of time, remaining at about 1 Pg C yr^{-1} from 2000 to 2100 (Fig. 3-3).

A *stabilizing policy* scenario (SP) was also developed, which assumes that deforestation stops by 2025 and that >1000 million ha are reforested by 2100. Under these scenarios, forests become a C sink by 2000 and gradually reach peak absorption of 0.7 Pg C yr^{-1} by 2025 (Fig. 3-3). This exercise suggests that a wide range of future emissions from deforestation is possible, and that the actual path that is taken will have a significant impact (positive or negative) on global warming.

Prospects for Reversing Deforestation

Halting deforestation and conducting reforestation may be one of the most cost-effective options for reducing net emissions of CO_2. Although the amount of land involved for any significant offset of CO_2 emissions is very large (Marland, 1988), the cost per Mg of C removed from the atmosphere can be very low compared to other control strategies (Dudek, 1988; Krause

& Koomey, 1989). Preliminary estimates suggest that it may be feasible to decrease clearing of forests and develop agressive reforestation programs that could reverse the current deforestation trend and perhaps significantly increase global forest biomass. (Postel & Heise, 1988; Meyers, 1989; Sedjo, 1989; Lashof & Tirpak, 1990).

Causes of deforestation are complex and vary from region to region. An effective strategy to reduce deforestation is therefore difficult to describe, but would include programs to both decrease the demand for wood, where harvesting is unsustainable, and to reverse economic incentives that currently encourage clearing. Demand for forest land and products can be reduced by substituting sustainable, sedentary agricultural technologies for shifting agriculture; improving the efficiency of biomass (fuelwood and charcoal) combustion in cookstoves; substituting durable products for disposable forest products; and recycling wood and paper products.

Increasing forest area and more intensive forest management can help to discourage the clearing of natural forests as well as increase total biomass. Opportunities include replanting marginal agricultural lands and degraded forests, improving management and productivity of existing forests; using agro-forestry techniques in the tropics; and establishing biomass-for-energy plantations. A combination of strategies using species optimized for rapid growth and for maximum standing biomass may be required.

Tree Planting on Surplus Agricultural Lands

Dudek (1988) has done a preliminary analysis of the costs of using some of the land in the U.S. Conservation Reserve Program (CRP) as a C sink. Costs may be as low as a few U.S. dollars per megagram of C removed, which is inexpensive compared with other CO_2 limiting strategies such as nuclear power, or scrubbing CO_2 from coal-fired power plants (Keepin & Kats, 1988; Krause & Koomey, 1989).

In the USA, trees planted on CRP land could offset additional CO_2 emissions from fossil fuel–burning power plants. For the 1987 to 1996 time period, an estimated 25 223 GW increase in fossil fuel–based electrical generation capacity would produce an additional 0.045 Pg of C yr^{-1} as CO_2 (NAERC, 1987 as cited in Dudek, 1988). Acreage required to achieve this offset will depend on the productivity of the trees that are planted. Assuming a relatively high net C-fixing rate (net ecosystem productivity) of 5 Mg C ha^{-1} yr^{-1} is achieved, only 9 million ha of land would be required. This area is about 50% of the 1990 CRP acreage goal of 16 to 18 million ha. Expansion of the land area planted in trees under the CRP could be achieved by requiring utilities to offset increased emissions and/or by increasing rental payments to farmers who plant trees. Assuming plantation establishment costs of \$430 ha^{-1} for planting and fertilization and rental payments of \$250 ha^{-1} yr^{-1} for 30 yr, establishment of 9 million ha would cost \$3.9 billion and rental payments would be \$2.2 billion yr^{-1}. Using a 10% discount rate and assuming that the trees continue to accumulate C for 30 yr, 0.045 Pg C yr^{-1} could be offset for a net present value of about \$25 billion or \$18.25 per Mg of C.

This rough calculation probably overstates the social costs of such a program because it does not take into account other benefits, such as soil conservation and marketable timber. Actual government outlays would be lower because rental payments will probably only last for 10 yr, and because only the marginal cost of getting farmers to plant trees, rather than another cover crop, should be charged to CO_2 removal (see Dudek, 1988).

Research Needs Relating to Land-Use Change

Research is needed both to improve emissions estimates and to refine response strategies. To improve estimates of emissions from land-use change, better estimates of initial standing biomass in the areas of clearing are needed. Better estimates or current rates of land clearing and amount of biomass burned, and improved emissions factors for biomass burning (emissions per mass burned) and land disturbance (total emissions per area disturbed) are also needed. To improve the prospect of reversing deforestation, research is needed into the social and economic factors leading to deforestation, the availability and characteristics of land that might be reforested, and the species and management techniques that could be used to increase forest productivity.

METHANE

Measurements of methane concentrations made in the 1980s show that the atmospheric burden of CH_4 has been increasing by about 1% yr^{-1} and had reached 1.7 $\mu mol\ mol^{-1}$ in 1987 (Blake & Rowland, 1988). Per molecule added to the atmosphere at current concentrations, the radiative forcing from CH_4 is about 20 times greater than from CO_2.

Methane is produced by anaerobic decomposition of organic matter. Sources include natural wetlands, rice paddies, biomass burning, domestic animals, landfills, and fossil fuel production. Total annual emissions of CH_4, inferred from atmospheric chemistry with an uncertainty of about 20%, are estimated to be about 500 Tg CH_4 yr^{-1} (1 Tg = 1 teragram = 10^{12} g) (Cicerone & Oremland, 1988). Although the major sources of CH_4 have been identified, there is considerable uncertainty regarding the contribution of each source. Of the global CH_4 budget, an estimated 60 to 170 Tg yr^{-1} are from rice, 65 to 100 Tg yr^{-1} are from domestic animals, and 50 to 100 Tg yr^{-1} are from biomass burning; other important sources include natural wetlands, landfills, coal production and natural gas production and distribution, and possibly termites[3] (Cicerone & Oremland, 1988) (Fig. 3-4).

The uncertainty of these estimates results from a lack of reliable emissions measurements for the range of conditions under which CH_4 is produced by each source. For example, no measurements of CH_4 emissions

[3] Termites of the geni *Reticulitermes, Cryptotermes, Coptotermes,* and *Gnathamitermes* produce methane. Examples of individual termite species that produce methone are *Reticulitermes tibielis* and *Gnathamitermes perplexus.*

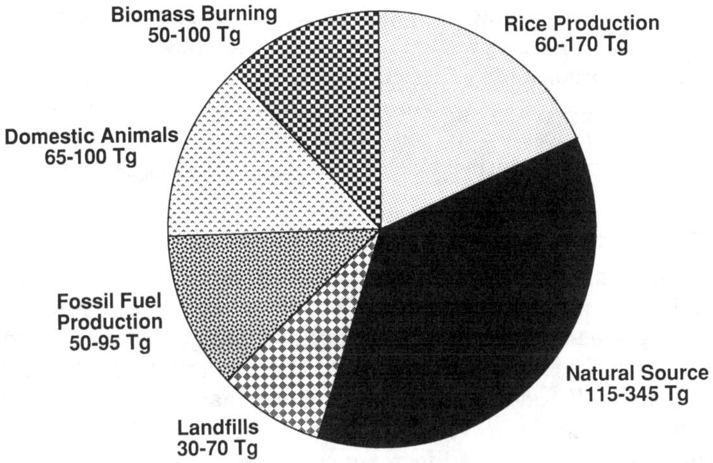

Fig. 3-4. Current global annual emissions of CH_4 by source. (teragrams CH_4 yr^{-1}). Human activities in the agricultural sector (domestic animals, rice production, and biomass burning) are major sources of atmospheric CH_4. Natural sources from wetlands, oceans, and lakes probably contribute less than 25% to the global CH_4 budget (sources: Cicerone & Oremland, 1988; Crutzen et al., 1986; Lerner et al., 1988; IRRI, 1986).

are available from the major rice-producing areas of Asia, where the majority of the world's rice is grown. Similarly, the available data do not characterize the full range of water-management regimes used in growing rice. For livestock, although many measurements have been made of emissions from large livestock in developed countries, better estimates are needed for a range of animals (including draft), on a range of forage diets in developing countries. As discussed above, biomass burning emissions are uncertain because both the amount burned and the emission factors are not well characterized.

Rice

Rice cultivation and productivity are increasing. Between 1950 and 1984, rough rice production nearly tripled, whereas the area under cultivation increased by only 40%. Global average yield doubled during this period from 1.6 to 3.2 Mg ha^{-1} (IRRI, 1986). China and India alone produce >50% of the world's rough rice.

Factors Affecting Emissions

Methane is produced by anaerobic decomposition of organic matter in flooded rice fields. Nonflooded, upland rice fields do not produce CH_4. The emissions rate is affected by the particular growth phase of the rice plant, temperature, irrigation and water management practice, fertilizer usage, presence of organic matter, rice species under cultivation, and number and duration of rice harvests (Cicerone et al., 1983; Seiler et al., 1984; Holzapfel-Pschorn & Seiler, 1986). The practice of incorporating crop residues into the rice paddy increases available organic material and subsequent production of methane. High-yielding varieties of rice, which have a higher grain-to-straw ratio than traditional varieties, produce less CH_4 per unit of rice. This improvement is due to the relative reduction in the amount of crop residue available for decomposition and the reduction in the maturation period, which reduces the time period during which the paddy is flooded (Dalrymple, 1986).

Strategies for Limiting Emissions

Methane emissions from rice production could be reduced through further productivity increases and through better management of crop residues, such as using a portion for animal fodder or building materials. Adoption of improved rice varieties with high yield potential, particularly those well suited to "upland" conditions would reduce methane emissions per unit of rice produced. Also, improved fertilization practices such as direct placement during transplant, could increase crop yield with no concomitant increase in CH_4.

Research Needs for Rice

Improved understanding of the factors controlling CH_4 emissions from rice cultivation is important, both for quantifying current sources and exploring potential approaches for controlling emissions. The influence of rice cultivar, cultivation practices, and water management regimes on CH_4 emissions need to be quantified. Further productivity increases would continue to reduce CH_4 emissions per unit of rice produced, and the development of high-yielding varieties for nonirrigated regimes, including uplands, would decrease the incentive to shift to high CH_4-emitting irrigated regimes. Research aimed directly at reducing CH_4 production in rice cultivation is also needed. The potential for developing new low-cost technology, such as new strains of rice, wind-driven devices to aerate paddy soils, and chemical inhibitors incorporated into fertilizer, should be explored.

Domestic Animals

Of the estimated 65 to 100 Tg of CH_4 that livestock contribute annually, approximately 57% is from cattle (*Bos* sp.), 19% is from dairy cows, and 10% is from sheep (*Ovis aries*) (Crutzen et al., 1986; Lerner et al., 1988) (Fig. 3–5). With the global cattle population increasing at $>1\%$ yr^{-1} (FAO, 1986), livestock will continue to be a growing source of CH_4. Livestock in India, USSR, Brazil, USA, and China produce nearly half of the CH_4 from this source.

Livestock systems vary considerably around the globe. In developed economies, livestock are intensively managed for either meat or dairy production. Within these systems livestock are typically larger animals, fed at least part of the time on a highly digestable grain diet. Yield of meat and/or dairy product are much higher than in developing countries, and correspondingly CH_4 emissions are higher per head, but lower per unit of animal product.

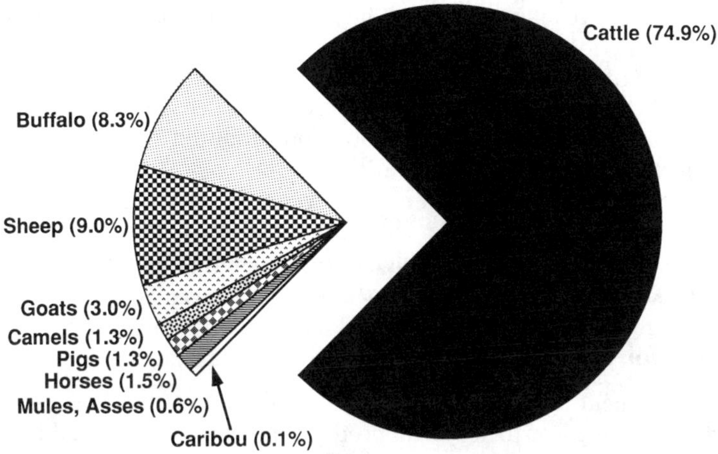

Fig. 3-5. Current annual global emissions of CH_4 from domestic animals by animal type. Of the estimated 65 to 100 Tg of CH_4 that domestic animals emit annually, the majority is from beef and cairy cattle, followed by sheep, buffalo (*Babalus babalus*), and goat (*Capra hircus*) (Camel, *camelus* spp.; pig, *Sus scrofa*; horse, *Equus caballus*; mule, *Equus caballus* × *Equus asinus*; caribou, *Rangifer* spp.) (source—Crutzen et al., 1986; Lerner et al., 1988).

In less developed economies, livestock are usually integrated into an overall crop/livestock or pasture-based system. These animals are typically smaller than in industrialized countries and live near the maintenance level on a low-quality forage diet. They are frequently scavengers and may add many benefits to the agricultural system. Beyond production of meat, dairy products, and fiber, livestock are a source of fertilizer, fuel, and traction power in many developing countries. They can also help to stabilize cash flows and food supplies when annual crops fail. Thus the contribution of livestock to the socio-economic systems in developing countries cannot be measured by meat and dairy production alone. Thus, it is difficult to evaluate CH_4 emissions per unit of "product" and to devise strategies for reducing CH_4 emissions.

Factors Affecting Emissions

The loss of feed energy in the form of CH_4 is an inefficiency in meat and dairy production. Estimates of energy loss as CH_4 expressed as a percentage of gross energy intake for cattle are in the range 3.5 to 8.3% (D. Johnson, 1988, personal communication; Blaxter & Wainman, 1964). Methane emissions from livestock are affected by differences in quantity and quality of feed, body weight, age, energy expenditure, and enteric ecology. Ruminants emit more CH_4 per unit of body weight than nonruminants, and all other things being equal, emissions are greater for heavier animals, higher feeding levels, and high cellulose forage diets. Emissions are also greater for working animals, because some of the additional feed requirement is converted to CH_4 (Lerner et al., 1988).

Feed quantity and composition interact in affecting CH_4 generation. Blaxter and Clapperton (1965) concluded that at low levels of feed intake—

at or slightly above maintenance—a high-cellulose forage diet results in lower CH_4-production per unit of feed than a high-N grain diet. Conversely, at high levels of feed intake—approaching three times maintenance—a high-quality grain diet results in lower CH_4 production per unit of feed. This suggests that current feeding practices in both industrialized and developing countries tend to minimize CH_4 generation.

Microbial decomposition in manure will result in the generation of CH_4 if conditions become anaerobic. Disposal of manure in lagoons and flooded fields produces conditions conducive to the generation of CH_4, but this source of CH_4 is not well quantified.

Strategies for Limiting Emissions

Management techniques to reduce emissions have not been widely explored. Increasing livestock system productivity and using feed additives can affect CH_4 generation, but uncertainties regarding emission factors makes development of a CH_4 reduction strategy difficult. In general, high productivity livestock systems, with animals fed at several times maintenance, produce less CH_4 per unit of animal product. Feed additives that increase feed use efficiency and reduce CH_4 generation by modifying rumen fermentation are currently available for beef cattle (Bergen & Bates, 1984). These feed additives, known as ionophores (carboxylic polyether ionophore antibiotics) are fed to most feedlot cattle to improve the efficiency of beef production during finishing for market. Ionophores improve the efficiency of beef production by 6 to 8% while reducing CH_4 production by approximately 5% for a forage diet and approximately 20% for a high quality grain diet (D. Johnson, 1988, personal communication).

Research Needs for Domestic Animals

In developing countries, estimates of CH_4 generation for a range of livestock (including draft animals) on a variety of diets is needed to provide the information necessary to devise a strategy to reduce CH_4 generation. Also, the potential for using antibiotics, such as ionophores, on cattle in developing countries needs to be explored. A systematic evaluation of interspecific and individual variations in CH_4 emissions might allow breeding to reduce emissions. Also, CH_4 emissions from manure are unquantified; CH_4 generation under a range of management and disposal options needs to be examined.

NITROUS OXIDE

Nitrous oxide is present in only small amounts in the atmosphere, but it can play an important role in both the radiation balance and the chemistry of the atmosphere. The current concentration of N_2O is 0.31 μmol mol^{-1}, three orders of magnitude less than CO_2, but its radiative forcing is 230 times greater per molecule added. Nitrous oxide can also contribute to stratospheric O_3 depletion by producing NO_x as it breaks up in the stratosphere. The annual growth rate of N_2O is currently about 0.2 to 0.3% yr^{-1} (Rasmussen & Khalil, 1986).

Nitrous oxide emissions result from fossil fuel combustion, biomass burning, nitrogenous fertilizer use, and land clearing, as well as natural processes in soils, oceans, and freshwater systems (Seiler & Conrad, 1987; Bolle et al., 1986). Total N_2O emissions can be inferred from atmospheric chemistry, but it is not clear whether currently identified sources can account for the observed rate at which its concentration is increasing (see below). The considerable uncertainty in the global N_2O budget and the contribution of individual sources arises from the scarcity of measurements of N_2O fluxes, the complexity of biogeochemical interactions that produce N_2O, and heterogeneity of the landscape where N_2O is produced.

Nitrous oxide is simultaneously produced and consumed in soils via the microbial metabolic pathways of denitrification, nitrification, nitrate assimilation, and nitrate dissimilation (Sahrawat & Keeney, 1986). Addition of nitrogenous fertilizer to the soil increases N_2O production, but, estimates of fertilizer induced N_2O emissions are highly variable (Galbally, 1985; Eichner, 1990). Recent measurements indicate that previous estimates of N_2O from fossil fuel combustion were much too high (Muzio & Kramlich, 1988). This suggests that other anthropogenic emissions, such as those related to land disturbance and fertilizer NO_3 leached into aquifers, may be underestimated in current budgets, or that there are additional sources that have not been identified. Of an estimated global annual N_2O source of 8 to 22 Tg yr^{-1}, nitrogenous fertilizer contributes approximately 0.14 to 2.4 Tg yr^{-1} (Lashof & Tirpak, 1990) and biomass burning contributes about 1 to 2 Tg yr^{-1} (Crutzen, 1983). As discussed above, exposed soils that result from deforestation and biomass burning may be a significant source of N_2O to the atmosphere. An estimate of the contribution from this source to the global budget has not been attempted.

Nitrogenous Fertilizer

Global N fertilizer consumption, 70.5 Tg N in 1984 to 1985 (FAO, 1986), is increasing at 1.3% yr^{-1} in industrialized countries and 4.1% yr^{-1} in developing countries (World Bank, 1988). Fertilizer is likely to become an increasingly significant source of N_2O in the future.

Factors Affecting Emissions

Practices affecting fertilizer-derived emission of N_2O include the fertilizer type, amount of fertilizer applied, application technique, timing of application, tillage practices, use of chemicals, irrigation practices, vegetation type, and residual N in the soil. Natural factors such as temperature, precipitation, soil organic matter content, and soil pH also affect N_2O emissions (Sahrawat & Keeney, 1986).

There is great uncertainty surrounding N_2O production as a function of fertilizer type. Estimates suggest that the fertilizer-derived emissions of N_2O are highest for anhydrous ammonia (between 0.5 and 6.8% of N applied), followed by urea (0.07–0.5% of N applied), ammonium nitrate, ammonium sulfate, and ammonium phosphate (0.03–1.7% of N applied), and N solutions (0.001–0.5% of N applied) (Eichner, 1990; Galbally, 1985). "Median" values given by Galbally (1985) are 0.5% for anhydrous ammonia, 0.1% for ammonium types, and 0.05% for nitrate fertilizer.

Nitrous oxide emissions are difficult to predict because of the complexity and variability of fertilized soil systems. Nitrous oxide emissions are likely to be higher with increased fertilizer application; broadcasting, rather than deep-placement of fertilizer; and when the soil is flooded intermittently, as in rice cultivation. Emissions may also be higher for fall, rather than spring application, and when herbicides are used. Although N_2O emissions have not been measured in rice paddies, denitrification losses of fertilizer are very high in these fields, suggesting that N_2O emission rates may also be high (Sing & Prasad, 1985).

Strategies for Limiting Emissions

Nitrous oxide emissions may be reduced by improving fertilizer use efficiency through management changes such as deep placement in the soil, or technological changes such as nitrification inhibitors, reduced-release-rate fertilizers, and fertilizer coatings. These practices increase the fraction of the applied fertilizer taken up by the plants and reduce total fertilizer requirements. Unfortunately, the impact of these techniques on N_2O emissions has not been measured.

Research Needs for Nitrogenous Fertilizer

Better quantification of the effects of various factors such as fertilizer type, crop type, application rate, fertilizer placement, water management practice, tillage practice, and cropping system on N_2O emissions is needed. Research is also needed to compare N_2O emissions due to use of commercial fertilizers with emissions due to alternative practices, such as legume intercropping and manuring.

CONCLUSIONS

Agriculture and land-use changes are significant sources of greenhouse gases, and if current trends persist, emissions will increase in the future. Sustainable, low-input sedentary agricultural practices and improved biomass utilization efficiency could reduce the rate of land-clearing and biomass burning in the tropics. At the same time, a program of global reforestation could partially offset fossil-fuel CO_2 emissions. Rice cultivation and livestock could be managed to reduce the production of CH_4. Fertilizer-use efficiency can be improved, producing higher yields, while at the same time presumably reducing the emission of N_2O to the atmosphere.

The uncertainties surrounding the emission factors for each of these agricultural activities makes development of an optimal greenhouse gas reduction strategy difficult. Research is needed to refine emissions estimates for each of these sources and to develop technologies and alternative management practices to reduce emissions. Understanding the current greenhouse gas emissions related to agriculture and land-use practices and developing options for reducing these emissions while increasing production are major challenges for the agricultural research community.

ACKNOWLEDGMENT

Research conducted by I. Fung, E. Matthews, R. Houghton, K. Andrasko, and B. Braatz for USEPA's Report to Congress, *Policy Options for Stabilizing Global Climate,* was invaluable in preparing this Chapter. We thank D. Johnson for providing unpublished data and useful discussions. We would also like to thank B. Kimball, N. Rosenberg, and L. Allen, Jr., for their very constructive reviews of the manuscript.

REFERENCES

Anderson, I., J. Levine, M. Poth, and P. Riggan. 1988. Enhanced biogenic emissions of nitric oxide and nitrous oxide following surface biomass burning. J. Geophys. Res. 93:3893–3898.

Andreae, M., E. Browell, M. Garstang, G. Gregory, R. Harriss, G. Hill, D. Jacob, M. Periera, G. Sachse, A. Setzer, P. Silva Dias, R. Talbot, A. Torres, and S. Wofsy. 1988. Biomass-burning emissions and associated haze layers over amazonia. J. Geophys. Res. 93:1509–1527.

Bergen, W., and D. Bates. 1984. Ionophores: Their effect on production efficiency and mode of action. J. Anim. Sci. 58:1465–1483.

Blake, D.R., and F.S. Rowland. 1988. Continuing worldwide increase in tropospheric methane, 1978 to 1987. Science (Washington, DC) 239:1129–1131.

Blaxter, K.L., and J.L. Claperton. 1965. Prediction of the amount of methane produced by runimants. Br. J. Nutr. 19:511.

Blaxter, K.L., and F.W. Wainman. 1964. Utilization of food by sheep and cattle. J. Agric. Sci. 57:419.

Bolle, H., W. Seiler, and B. Bolin. 1986. Other greenhouse gases and aerosols: Assessing their role for atmospheric radiative transfer. p. 157–203. *In* B. Bolin et al. (ed.) The greenhouse effect, climate change, and ecosystems. John Wiley & Sons, New York.

Bowden, W.B., and F.H. Bormann. 1986. Transport and loss of nitrous oxide in soil water after forest clear-cutting. Science (Washington, DC) 233:867–869.

Cicerone, R., J. Shetter, and C. Delwiche. 1983. Seasonal variations of methane flux from a California rice paddy. J. Geophys. Res. 88:11022–1024.

Cicerone, R.J., and R.S. Oremland. 1988. Biogeochemical aspects of atmospheric methane. Global Biogeochem. Cycles 2:299–327.

Crutzen, P.J. 1983. Atmospheric interactions—homogenous gas reactions of C, N, and S containing compounds. p. 67–114. *In* B. Bolin and R.B. Cook (ed.) The major biogeochemical cycles and their interactions, SCOPE 21. John Wiley & Sons, New York.

Crutzen, P., I. Aselmann, and W. Seiler. 1986. Methane production by domestic animals, wild runimants, other herbivorous fauna, and humans. Tellus Ser. B 38:271–284.

Crutzen, P., L. Heidt, J. Krasnec, and W. Pollock. 1979. Biomass burning as a source of atmospheric gases CO, H_2, N_2O CH_3Cl, and CO_2. Nature (London) 282:253–256.

Dalrymple, D. 1986. Development and spread of high-yielding rice varieties in developing countries. Agency for Int. Dev., Washington, DC.

Detwiler, R.P., and C.A.S. Hall. 1988. Tropical forests and the global carbon cycle. Science (Washington, DC) 239:42–50.

Dudek, D. 1988. Offsetting new CO_2 emissions. Rep. of the Environ. Defense Fund, New York.

Eichner, M. 1990. Nitrous oxide emissions from fertilized soils: Summary of available data. J. Environ. Qual. 19:272–280.

Food and Agriculture Organization of the United Nations. 1986. 1985 FAO production yearbook. Vol. 39. FAO, Rome, Italy.

Friedli, H., H. Lotscher, H. Oeschger, U. Siegenthaler, and B. Stauffer. 1986. Ice core record of the $^{13}C/^{12}C$ ratio of atmospheric carbon dioxide in the past two centuries. Nature (London) 324:237.

Fung, I. 1986. Analyses of seasonal and geographical patterns of atmospheric CO_2 distributions with a three-dimensional tracer model. p. 459–473. *In* J.R. Trabalka and D.E. Reichle (ed.) The changing carbon cycle: A global analysis. Springer-Verlag New York, New York.

Galbally, I.E. 1985. The emission of nitrogen to the remote atmosphere: Background paper. p. 27–53. *In* J.N. Galloway et al. (ed.) The biogeochemical cycling of sulfur and nitrogen in the remote atmosphere. Reidel, Dordrecht, Netherlands.

Hansen, J., I. Fung, A. Lacis, S. Lebedeff, D. Rind, R. Ruedy, G. Russell, and P. Stone. 1988. Global climate change as forecast by the GISS 3-D model. J. Geophys. Res. 93:9341–9364.

Holzapfel-Pschorn, A., and W. Seiler. 1986. Methane emission during a cultivation period from an Italian rice paddy. J. Geophys. Res. 91:11 803–11 814.

Houghton, R., J. Hobbie, and J. Melillo. 1983. Changes in the carbon content of terrestrial biota and soils between 1860 and 1980: A net release of CO_2 to the atmosphere. Ecol. Monogr. 53:235–262.

Houghton, R., R. Boone, J. Fruci, J. Hobbie, J. Melillo, C. Palm, B. Peterson, G. Shaver, G. Woodwell, B. Moore, D. Skole, and N. Myers. 1987. The flux of carbon from terrestrial ecosystems to the atmosphere in 1980 due to changes in land use: Geographic distribution of the global flux. Tellus Ser. B 39:122–139.

International Rice Research Institute. 1986. World rice statistics 1985. IRRI, Manilla, Philippines.

Keeling, C.D., and M. Heimann. 1986. Meridional eddy diffusion model of the transport of atmospheric carbon dioxide. 2, Mean annual carbon cycle. J. Geophys. Res. 91:7782–7796.

Keepin, B., and G. Kats. 1988. Greenhouse warming: Comparative analysis of nuclear and efficiency abatement strategies. Energy Policy 16:538–561.

Krause, F., and J. Koomey. 1989. Unit cost of carbon savings from urban trees, rural trees, and electricity conservation: A utility cost perspective. *In* K. Garbesi et al. (ed.) Proc. of the Heat Islands Workshop, Berkeley, CA 23–24 February. LBL Publ. 27872. Lawrence Berkeley Lab., Energy Analysis Program, Berkeley, CA.

Lashof, D.A., and D. Tirpak. 1990. Policy options for stabilizing global climate. USEPA, Washington, DC.

Lerner, J., E. Matthews, and I. Fung. 1988. Methane emission from animals: A global high-resolution database. Global Biogeochem. Cycles 2:139–156.

Logan, J. 1983. Nitrogen oxide in the troposphere: Global and regional budgets. J. Geophys. Res. 88:10 785–10 807.

Marland, G. 1988. The prospect of solving the CO_2 problem through global reforestation—TR039. DOE/NBB-0082. U.S. Dep. of Energy, Washington, DC. Available from NTIS, Sprinfield, VA.

Meyers, N. 1989. The greenhouse effect: A tropical forestry response. Biomass 18:73–78.

Muzio, L.J., and J.C. Kramlich. 1988. An artifact in the measurement of N_2O from combustion sources. Geophys. Res. Lett. 15:1369–1372.

Neftel, A., E. Moor, H. Oeschger, and B. Stauffer. 1985. Evidence from polar ice cores for the increase in atmospheric CO_2 in the past two centuries. Nature (London) 315:45–47.

North American Electric Reliability Council. 1987. Electricity supply and demand for 1987–1996. NAERC, Princeton, NJ.

Postel, S., and L. Heise. 1988. Reforesting the Earth. Worldwatch Pap. 83. Worldwatch Inst., Washington, DC.

Ramanathan, V., R.J. Cicerone, H.B. Singh, and J.T. Kiel. 1985. Trace gas trends and their potential role in climate change. J. Geophys. Res. 90:5547–5566.

Rasmussen, R., and M. Khalil. 1986. Atmospheric trace gases; trends and distributions over the last decade. Science (Washington, DC) 232:1632–1634.

Sahrawat, K., and D. Keeney. 1986. Nitrous oxide emission from soils. Adv. Sci. 4:103–148.

Sedjo, R.A. 1989. Forests: A tool to moderate global warming? Environment 31:14–20.

Seiler, W., and R. Conrad. 1987. Contribution of tropical ecosystems to the global budgets of trace gases, especially CH_4, H_2, CO and N_2O. p. 133–162. *In* R. Dickinson (ed.) The geophysiology of Amazonia: Vegetation and climate interactions. John Wiley & Sons, New York.

Seiler, W., A. Holzapfel-Pschorn, R. Conrad, and D. Scharffe. 1984. Methane emission from rice paddies. J. Atmos. Chem. 1:241–268.

Siegenthaler, V., and H. Oeschger. 1987. Biospheric CO_2 emissions during the past 200 years reconstructed by deconvolution of ice core data. Tellus Ser. B 39:140–154.

Singh, S., and R. Prasad. 1985. Studies on the nitrification inhibitor dicandiamide (DCD) for increasing the efficiency of nitrogen applied to rice. J. Agric. Sci. 104:425–428.

Smith, J., and D. Tirpak. 1989. Potential effects of global climate change on the United States. USEPA, Washington, DC.

Stevens, C., and A. Engelkemeir. 1988. Stable carbon isotopic compositions of methane from some natural and anthropogenic sources. J. Geophys. Res. 93:725–733.

Tans, P., T. Conway, and T. Nakazawa. 1989. Latitudinal distribution of the sources and sinks of atmospheric carbon dioxide derived from surface observations and an atmospheric transport model. J. Geophys. Res. 94:5151–5172.

Tucker, C., J. Townshend, T. Goff, and B. Holben. 1986. Continental and global scale remote sensing of land cover. p. 221–241. *In* J.R. Trabalka, and D.E. Reichle (ed.) The changing carbon cycle: A global analysis. Springer-Verlag New York, New York.

Woodwell, G., R. Houghton, T. Stone, and A. Park. 1986. Changes in the area of forests in Rondonia, Amazon Basin, measured by satellite imagery. p. 242–257. *In* J.R. Trabalka, and D.R. Reichle (ed.) The changing carbon cycle: A global analysis. Springer-Verlag New York, New York.

World Bank. 1988. World Bank, FAO, UNIDO Fertilizer Working Group. Nitrogen supply, demand and balances for 1986/87 to 1992/93. World Bank, Washington, DC.

World Meteorological Organization. 1985. Atmospheric ozone 1985. World Meteorol. Org., Global Ozone Res. and Monitoring Project, Rep. 16. WMO, Geneva.

4 Effects of Carbon Dioxide on Photosynthesis, Plant Growth, and Other Processes[1]

Basil Acock

Systems Research Laboratory
Beltsville, Maryland

ABSTRACT

The increasing concentrations of CO_2 and other "greenhouse gases" are expected to change temperature, rainfall, and cloudiness. Since the late 17th century we have been collecting empirical information on crop response to all these climatic factors except CO_2 concentration ([CO_2]). The primary effects of high [CO_2] are an increase in photosynthetic rate and a decrease in transpiration rate. The former response is more pronounced in C_3 plants than in C_4 plants. The extra C fixed increases the dry weight of all organs, with disproportionately more going to roots and stems in many plants. The sizes of most vegetative organs and the numbers of many organs increase unless limited by other environmental factors. The environmental factors that can limit plant response to high [CO_2] include nutrients, water, temperature, and light, although the effects of some of these are ameliorated by high [CO_2]. These factors and [CO_2] interact in a complex way, so it is necessary to understand the mechanism of plant response to [CO_2] to predict growth and yield. The net result is that in a future high-[CO_2] world, we can expect larger plants and higher yields whatever organs are harvested, with higher water use efficiency but higher fertilizer requirements. The C_3 crops might displace maize (*Zea mays* L.) where water is adequate, because their yields will probably be stimulated more. However, the increased [CO_2] might make maize more drought-tolerant.

Earlier chapters in this book have documented how the increasing concentrations of CO_2 and various trace gases (the greenhouse gases) are expected to change temperature, rainfall, and cloud cover, i.e., light flux at the Earth's surface. Since the late 17th century we have been collecting empirical data on plant responses to temperature, rainfall, and light, but we have much less information about plant responses to CO_2 concentration ([CO_2]). In trying

[1]Contribution from the USDA-ARS.

 Impact of Carbon Dioxide, Trace Gases, and Climate Change on Global Agriculture, ASA Special Publication no. 53.

to predict the impact of climate change on agriculture, our fund of empirical knowledge would stand us in good stead were it not for the CO_2 factor. Therefore, this chapter will concentrate on plant responses to CO_2.

Because there have been several recent attempts to review and summarize the literature on plant response to CO_2 (e.g., Lemon, 1983; Strain & Cure, 1985; Enoch & Kimball, 1986), I have decided to concentrate on plant physiology and try to give a modeler's insight into why plants may be responding the way they do to [CO_2].

In 1978, Congress passed the National Climate Program Act, which charged the Department of Energy (DOE) with finding out what the increasing global [CO_2] would do to society. The DOE responded by setting up a Carbon Dioxide Research Program to answer this question. As of 1989, the work done under this program on agronomic crops alone had resulted in at least 73 peer-reviewed articles, 42 technical reports, 68 abstracts of talks and poster presentations, 9 book chapters, and 4 dissertations. Progress is summarized annually in a report "Carbon dioxide and climate: Summaries of research in 19XX" available from the DOE. Most of what we know about plant response to [CO_2] during long-term exposure comes from this research program.

THE HIERARCHY OF CARBON DIOXIDE CONCENTRATION EFFECTS

The most obvious effects of [CO_2] on plant photosynthesis and growth are shown in Fig. 4–1. They are arranged to show the hierarchy of primary, secondary, and lower order effects. First, there is an increase in photosynthesis and a reduction in photorespiration, which increases C availability to the

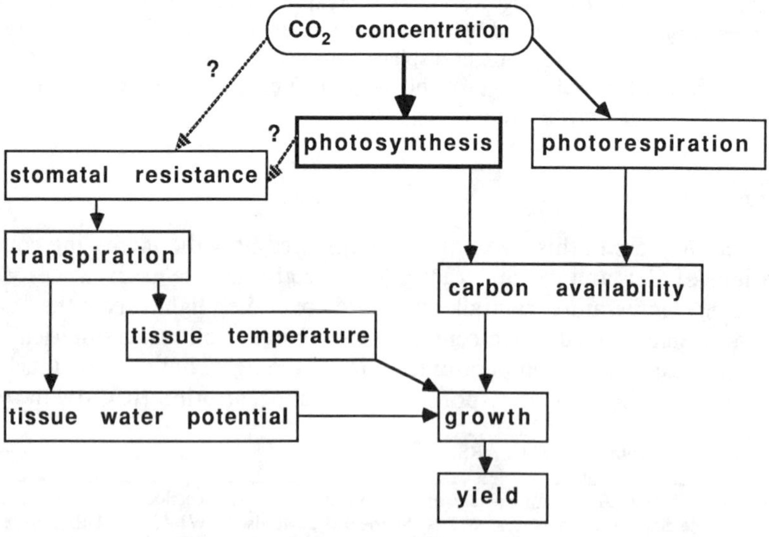

Fig. 4–1. A hierarchy of plant responses to [CO_2].

plant, and this, in turn, increases growth and yield. There is also an effect of [CO_2] on stomatal resistance and we don't really understand the mechanism for this effect. We are not sure how to model stomatal resistance, even after researching it for several decades. Stomata appear to respond directly to [CO_2], but that response is moderated by the photosynthetic characteristics of the plant (Farquhar & Sharkey, 1982). Whatever the mechanism, in high [CO_2] stomatal resistance is increased and transpiration rate per unit leaf area is decreased. This increase in stomatal resistance generally has the effect of raising tissue temperature and can also improve the plant tissue water relations. Both of these effects can increase growth and yield.

PHOTOSYNTHETIC CHEMISTRY

One reason why an increase in [CO_2] increases photosynthetic rate is that the gradient in [CO_2] between the outside and inside of the leaf is increased. Assuming that there is no change in stomatal resistance, the steeper gradient will increase the flux of CO_2 into the leaf and therefore will increase the photosynthetic rate. However, there are very interesting differences between various groups of plants. In Fig. 4–2 leaf net photosynthetic rate is plotted against [CO_2] for maize and wheat (*Triticum aestivum* L.) (Akita & Moss, 1973). The wheat plants go on responding up until 800 μmol mol^{-1} and perhaps beyond, whereas the maize does not respond very much after 400 μmol mol^{-1}. The difference between them is that wheat belongs to a group of plants called the C_3 plants, which include most agronomic plants; maize belongs to another group of plants called the C_4 plants, which includes the tropical grasses. At the 1990 [CO_2], maize has a photosynthetic advantage over wheat, and consequently, a greater yield. This is probably

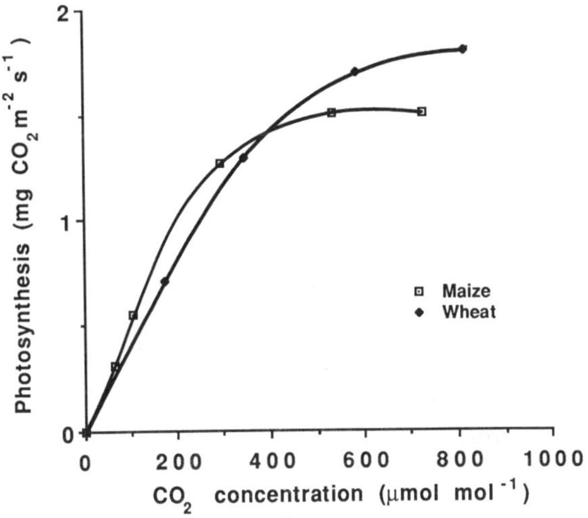

Fig. 4–2. Leaf net photosynthetic rate as a function of [CO_2] for maize and wheat (after Akita & Moss, 1973).

why so much maize is grown. But, about the turn of the century, if [CO_2] reaches the concentrations projected, maize will probably lose that advantage, and it will be interesting to see how much of it is grown after that. What is the reason for this difference between C_3 and C_4 plants?

The reactions that take place in the chloroplasts are shown in Fig. 4–3. In the light reaction, energy from light effectively splits water to release oxygen and provide a source of chemical energy (ATP) and a reducing compound ($NADPH_2$). The ATP and $NADPH_2$ are used in the dark reaction to reduce CO_2 to sugars (Fig. 4–3). The C_3 and C_4 groups of plants differ in their dark reactions.

Figure 4–4 shows the principal dark reactions that are possible in plants (Beer, 1986). The reactions are grouped into two compartments, because in different species, these reactions occur in different places. In the lower half

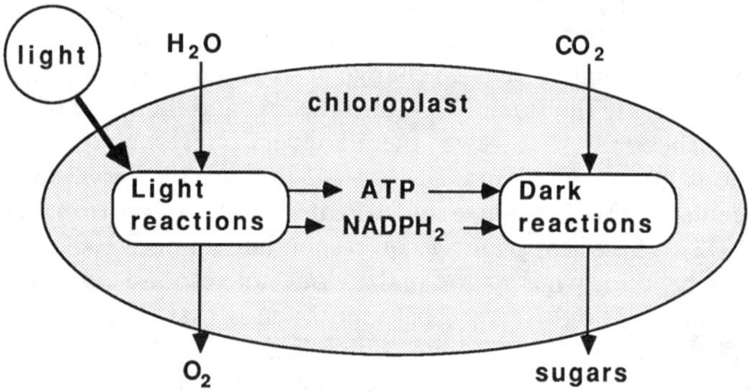

Fig. 4–3. Basic photosynthetic reactions occurring in the chloroplasts of leaves. ATP is a source of energy and $NADPH_2$ is a reducing compound.

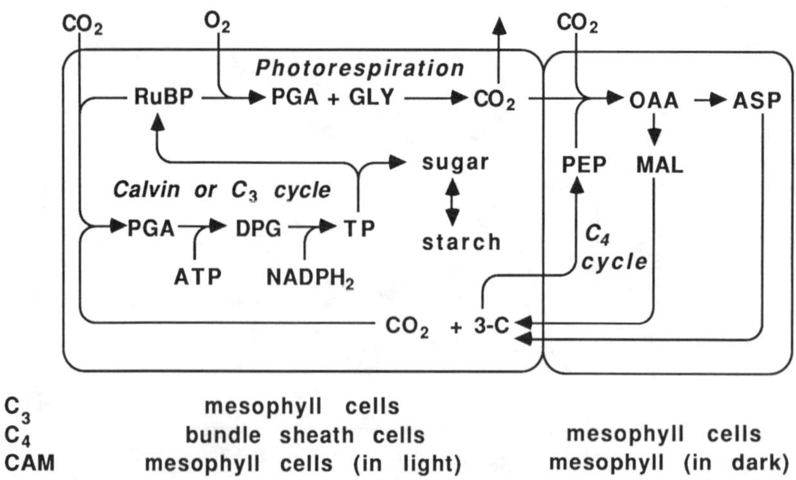

Fig. 4–4. The chemistry and location of the principal dark reactions in C_3, C_4, and CAM plants. For the names of compounds involved in these reactions refer to any book on photosynthetic chemistry.

of the left-hand compartment, the ATP energy source and the $NADPH_2$ reducing compound feed into the Calvin or C_3 cycle. It is called the *C_3 cycle* because C is cycled through a set of reactions mostly involving compounds with three C atoms. In this Calvin cycle, CO_2 is reduced to sugar and starch, which the plant uses to grow. However, during this cycle, ribulose bisphosphate (RuBP) is produced; RuBP can react either with CO_2 or oxygen (O_2). If it reacts with O_2, then C is lost from the cycle in a reaction called photorespiration, and less sugar and starch are produced. On the other hand, if [CO_2] is high or [O_2] is low, relatively more of the RuBP reacts with CO_2, and photorespiration is reduced, whereas sugar and starch production is increased. In the right-hand compartment, CO_2 is temporarily fixed and then released again in the C_4 cycle. Yes, the compounds involved in the *C_4 cycle* mostly have four C atoms.

All green plants are capable of doing the chemistry in the left-hand compartment of Fig. 4–4, and C_3 plants are limited to that photosynthetic chemistry. Underneath the compartments in Fig. 4–4, the locations of the chemical reactions are shown for the three major photosynthetic groups of plants. In C_3 plants, the Calvin cycle occurs in mesophyll cells. These are thin-walled cells exposed to the air inside the leaf so CO_2 and O_2 readily diffuse in and out of the cells. As mentioned before, the RuBP produced in the Calvin cycle can either react with CO_2 or O_2. As a result, in C_3 plants, increasing the [CO_2] in the air around the plants, increases the amount of CO_2 competing with O_2 for RuBP, and consequently increases the amount of activity in the Calvin cycle at the expense of photorespiration.

In C_4 plants, the reactions in the left-hand compartment occur in the bundle sheath cells, which are separated from the air inside the left by several layers of other cells. As a result, CO_2 and O_2 only diffuse very slowly to and from the bundle sheath cells. However, C_4 plants also have the C_4 cycle, which fixes CO_2 in the mesophyll cells and then releases it in the bundle sheath cells. In this way, C_4 plants supply CO_2 to the Calvin cycle while reducing the access of O_2. Thus, very little photorespiration occurs in C_4 plants. These differences in photosynthetic chemistry explain the responses seen in Fig. 4–2. The C_4 plants reduce photorespiration and this gives them an advantage over C_3 plants in low [CO_2]. However, C_4 plants fix CO_2 by a less direct method than C_3 plants, which puts C_4 plants at a disadvantage in high [CO_2] when photorespiration in C_3 plants is suppressed. Patterson and Flint (Chapter 7, this book) describe several experiments in which C_3 plants growing in competition with C_4 plants gained a distinct advantage at elevated [CO_2].

Another type of photosynthetic chemistry called *crassulacean acid metabolism* (CAM) occurs in cacti and succulents. They can perform all the chemistry in Fig. 4–4, and all the reactions take place in the mesophyll cells. However, the reactions are separated in time rather than in space, as in C_4 plants. The reactions in the right-hand compartment take place in the dark while the stomata are open; the reactions in the left-hand compartment take place in the light while the stomata are closed. This separation of the reactions in time allows the plants to conserve water. The CAM plants include pineapple (*Ananas comasus* Merr.) and kalanchoe (*Kalanchoe* spp.).

More detailed information about the photosynthetic chemistry of C_3, C_4, and CAM plants in relation to [CO_2] may be found in Tolbert and Zelitch (1983).

UNDERSTANDING CARBON PARTITIONING

How does the plant use this extra C? How is it partitioned? Table 4–1 shows for soybean [*Glycine max* (L.) Merr.] and maize, the percentage increase in dry weight for various organs on plants grown in high [CO_2] compared with their dry weight on plants grown in 340 μmol mol^{-1} [CO_2]. In both species, there is a large increase in the proportion of the dry weight, which goes to roots, a smaller increase to stems, and the smallest increase to leaves. As we have seen already in the photosynthesis data, there is a response to [CO_2] >520 μmol mol^{-1} in soybean but very little response in maize. What we saw in terms of photosynthesis is borne out in terms of dry matter increase.

Can we understand why the plant is partitioning in this way? I believe that we can if we accept teleological rather than mechanistic explanations. There are two principles guiding partitioning. The first is that, in the course of evolution, plants developed priorities for C use that ensured their survival. Therefore, the plants' top priority is probably survival of genes, followed by reproduction, growth of existing organs, increasing the number of organs (mostly by branching), storage of excess C, and, as excess C accumulates, maybe a reduction in C fixation rate.

What does survival of genes mean? It means different things to different plants. To a seed-bearing annual, presumably it means seed production. To a perennial, presumably it means the survival of the part of the plant that will grow again after a period of stress. Of course, the stage of development of the plant and environmental limitations may prevent the plant's addressing some of these priorities. If an annual plant has not reached the flowering stage, it cannot allocate C to reproduction and so will address lower priorities. If a plant is at a very low temperature, it may not be able to grow existing organs, so it moves on down the priority list.

Instead of thinking of the plant as having the same set of priorities throughout its life, it is equally valid to conceive of its having different sets of priorities at various growth stages. However, it probably always has those priorities in the order given above.

The second principle that helps us understand partitioning is that the plant has to maintain functional equilibrium (Thornley, 1972; Szaniawski, 1983). It has to balance the rates at which it takes in water and C and nutrients so that they are available in the correct ratio for growth. This means that during stress, or a shortage of one of these materials, the organs responsible for acquiring that material will get more C. For example, in a period of water stress, a plant will allocate proportionately more C to root growth than to shoot growth.

Table 4-1. Dry weights of organs on 11-wk old soybean and maize plants grown in various [CO_2]. The data are expressed as a percentage increase over those on plants grown in 340 μmol mol^{-1} (Rogers et al., 1983).

Plant part	CO_2 conc., μmol mol^{-1} 520	718	910
	% increase		
Soybean			
Roots	96	106	151
Stems	69	110	115
Leaves	40	65	86
Total shoot	52	75	98
Maize			
Roots	114	68	104
Stems	34	27	46
Leaves	15	8	18
Total shoot	34	34	39

It is actually possible to construct a complete partitioning scheme based on these two guiding principles. Such schemes can be used in plant models or to just help us formulate and test hypotheses about what plants are doing. The scheme in Fig. 4-5 will not be acceptable to everyone but may provide a framework that others can modify to suit their own understanding. In Fig. 4-5, C translocated from the leaves is used first to grow seeds and fruits. Anything remaining, depending on the N supply and demand, will go either to the organs that acquire N or to the rest of the plant. In a leguminous plant, the organs that acquire N are nodules; in nonleguminous plants, the roots acquire N. The C going to the rest of the plant is again

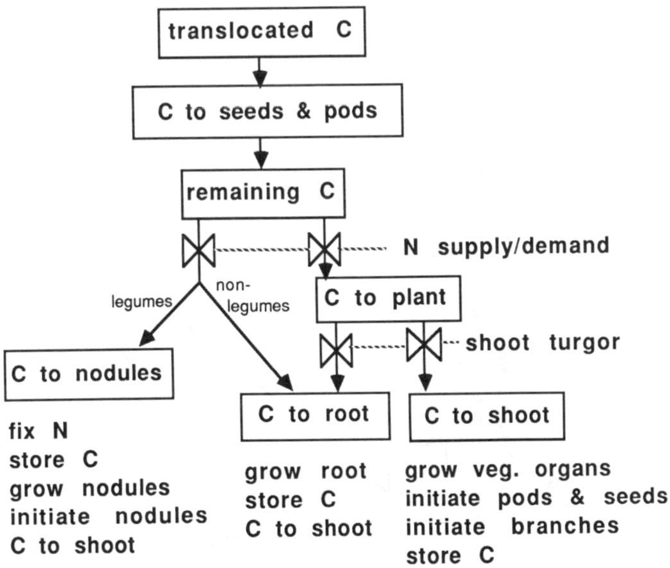

Fig. 4-5. A hypothetical scheme for partitioning C between organs on plants.

partitioned between roots and shoots, depending on shoot turgor. For each group of organs on the plant there are priorities for use of the C partitioned to them. From this scheme it is apparent that putting a plant in high [CO_2] and thereby increasing the amount of C entering the system will increase the need for N and hence the mass of roots, nodules, or both.

HOW PLANT ORGANS USE THE EXTRA CARBON

We have established that the mass of all organs on the plant will increase in high [CO_2]. But will the increased mass result in an increase in size or density of the organs? In general, tissue expansion is controlled by temperature and water relations, although nutrients and other factors may be limiting under certain circumstances. Tissue dry weight is controlled mainly by C availability. For a given volume of tissue, there are upper and lower limits to the amount of dry matter the tissue can contain. As a result, the relationship between tissue density (dry wt./vol.), the rate of new tissue formation, and C availability for all organs is something like that shown in Fig. 4–6. There is a certain minimum dry weight in the structure needed to support a volume of tissue. So if C availability is very low, it may actually reduce the volume of tissue that the plant can make. If C availability increases, as it will with high [CO_2], then tissue density increases until ultimately the plant cannot pack any more dry matter into the volume available. These two points of minimum and maximum density may be closer together for some tissues than others. If they are fairly close together, as in many seeds, then tissue density will be insensitive to [CO_2]. If they are far apart, as in leaves, tissue density will increase as [CO_2] increases. For example, the leaves of soybean

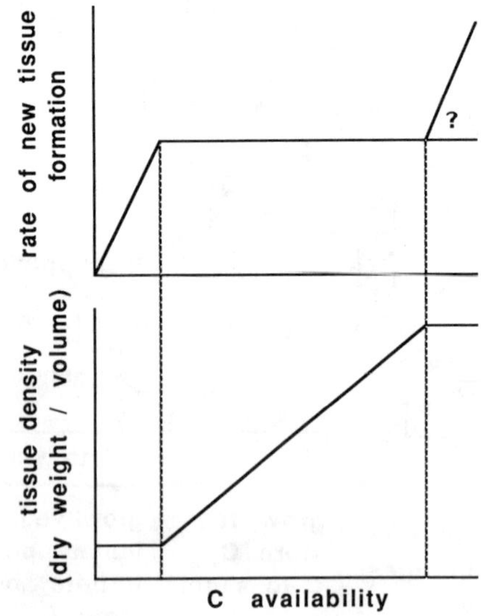

Fig. 4–6. A hypothetical relationship between tissue density (dry wt./vol.), the rate of new tissue formation, and C availability.

plants grown in 800 μmol mol^{-1} [CO_2] are typically 50% denser than those on plants grown in 330 μmol mol^{-1} [CO_2] (Acock & Allen, 1985).

Is the increase in density in high [CO_2] merely caused by the accumulation of starch and sugar in the leaf? It is possible. Ho (1978) found for tomato (*Lycopersicon esculentum* Mill.) that 65% of the C fixed is translocated immediately and 35% is stored as starch in the leaf. At the same time 12.5% of the stored starch is mobilized and translocated each hour. Increasing the C fixation rate in high [CO_2] increases the amount of starch stored in the leaf, and this alone changes tissue density. However, Fig. 4–7 shows data for the dry weights of structural matter and stored carbohydrates per unit leaf area for some melon (*Citrullus* sp.) plants. Clearly, the structural component contributes to the changes in specific leaf area, although this can be completely overshadowed by the change in stored carbohydrates. In these data, the structural component of leaf dry weight responded to [CO_2] in the same way at both 20 and 40 °C day temperature, but the storage component was larger at 20 °C, especially with higher [CO_2].

Some plants grown in high [CO_2] adapt to the increased C availability by increasing tissue volume. In some leaves, this increase in volume is achieved by adding layers of mesophyll cells, especially palisade mesophyll. For example, in soybean leaves grown in 910 μmol mol^{-1} [CO_2], the palisade layer is 53% thicker than in leaves grown in 340 μmol mol^{-1} [CO_2] (Thomas & Harvey, 1983). This additional volume may explain the increase in structural dry matter per unit leaf area shown in Fig. 4–7. Generally, the size of individual mesophyll cells is unchanged (Thomas & Harvey, 1983), and leaf area responds only slightly to [CO_2] (Acock & Allen, 1985). So, soybean leaves respond to high [CO_2] with increases in both tissue volume (mostly thickness) and density, and this is true of many species. Stem tissues may become denser or increase in volume depending on the species (Rogers et al., 1983) and roots do both (Del Castillo et al., 1989).

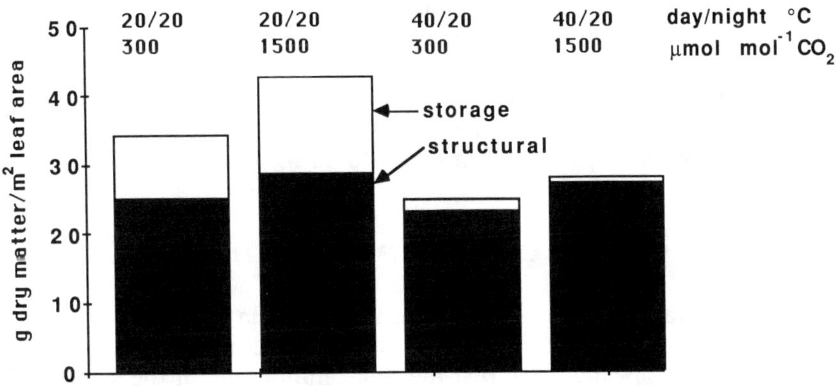

Fig. 4–7. Dry weight of structural material and stored carbohydrates in the leaves of 46-day-old melon plants grown in various temperatures and [CO_2] (Acock & Pasternak, 1986).

INITIATION AND MAINTENANCE OF NEW GROWING POINTS

When we compare plants grown in, say, 330 μmol mol^{-1} and 600 μmol mol^{-1} [CO_2], we normally see a small difference in the leaf area and stem volume, which suggests the growth of these organs is only slightly C-limited in 330 μmol mol^{-1} [CO_2], and some other factor is mainly responsible for limiting tissue expansion (Acock & Allen, 1985). The plants from 600 μmol mol^{-1} [CO_2] have higher C availability, and therefore, denser leaf and stem tissues. There is often C left over after these tissues have been filled to their maximum density, however, and the plant uses this excess C to start or maintain new growing points. During the vegetative phase, this means new branches on shoots and roots; during the reproductive phase, it means filling more of the fruits and seeds. This phenomenon can easily be observed for the shoot and has also been reported for roots.

Del Castillo et al. (1989) measured root elongation against the glass face of a rhizotron for well-watered soybean plants grown in a range of [CO_2]. At the end of the season, the accumulated root length was proportional to the [CO_2] in which the plants grew. Root elongation rate, measured on individual root axes, was not significantly affected by [CO_2], but the number of actively growing root tips increased with increasing [CO_2]. In this study, on well-watered plants grown in high [CO_2], the roots branched more and explored a given volume of soil more thoroughly, but they did not explore more soil with the extra root growth. The results might, of course, be different for other species and soil conditions.

REDUCTION IN PHOTOSYNTHETIC RATE

What happens when the plant is able to fix more C than it can use in growth? Micrographs of leaf tissue grown in high [CO_2] show large starch grains crushing the grana in the chloroplasts of mesophyll cells (e.g., Cave et al., 1981). The physical disruption of the photosynthetic system alone might explain why some plants have reduced photosynthetic rate when there are large accumulations of starch in the leaves. There are also biochemical pathway explanations of why photosynthesis might be reduced when this sort of thing occurs (e.g., Hesketh et al., 1983), but the truth is that we don't yet understand the phenomenon. This is another of the important unknowns in determining how plants will respond to an increase in [CO_2].

PLANT AND CANOPY TRANSPIRATION RATE

At the start of this chapter I mentioned that in high [CO_2], stomatal resistance increases, and transpiration rate per unit leaf area decreases. This does not automatically mean that the transpiration rate of the whole plant decreases in high [CO_2]. What happens on a canopy or land-area basis depends on how much bigger the plant is in the high [CO_2]. If the plant is

Table 4–2. Percentage contribution of changes in net photosynthesis (PN) and transpiration rate (TR) to a CO_2-induced doubling of leaf water use efficiency for several species (Acock & Allen, 1985).

Crop	PN	TR
Maize	27	73
Cotton†	60	40
Snow gum‡	75	25
Soybean	90	10

† *Gossypium hirsutum* L. ‡ *Eucalyptus* sp.

much bigger because of increased leaf area or by a virtue of branching, then there is more leaf area transpiring. So, canopy transpiration rate depends very much on how the increased leaf area and the decreased leaf transpiration rate balance out. There are results in the literature that show both an increase and a decrease in canopy transpiration rate in high [CO_2] (see references cited by Acock & Allen, 1985).

WATER USE EFFICIENCY

Even if canopy transpiration rate increases in high [CO_2], there is usually an increase in water use efficiency. This is the amount of C or dry matter accumulated per unit of water taken up or lost by the plant. It is instructive to look at how much of that change in water use efficiency comes from changes in photosynthetic rate and how much of it comes from changes in transpiration rate. Table 4–2 shows data for various crops where an increase in CO_2 concentration has doubled the rate of leaf water use efficiency (WUE). In maize, only 27% of the change in WUE can be attributed to an increase in photosynthetic rate, whereas 73% of the change is attributable to a decrease in transpiration rate. This difference is expected because we saw in Fig. 4–2 that the photosynthetic rate of maize does not respond greatly to [CO_2]. Instead, maize responds to high [CO_2] by increasing its stomatal resistance. I suggested earlier that maize will lose its photosynthetic advantage over C_3 crops around the year 2000, and we may see less maize grown. However, we may still see maize grown for its drought tolerance, caused by an increased stomatal resistance and decreased transpiration rate in high [CO_2]. At the bottom of the list in Table 2, soybean gets almost all of its change in WUE from the increase in photosynthetic rate and very little from closing its stomata. Other plants in the table are intermediate in their response.

NITROGEN FIXATION IN LEGUMES

In the earlier section on partitioning, I mentioned that the larger dry weight of plants grown in high [CO_2] requires them to acquire more N. For legumes, there is considerable evidence that they take up any fertilizer N available in the soil and then develop nodules to fix from the air any further N needed. If plants are grown in various [CO_2] but in the same soil, the plants in high [CO_2], being larger, have to fix proportionately more of their N_2. This is all the explanation needed for the stimulation of N_2 fixation in high [CO_2]. I do not know of any data that demands a direct effect of [CO_2] on N_2 fixation.

PREDICTING PLANT GROWTH AND YIELD IN FUTURE CLIMATES WITH HIGH CARBON DIOXIDE CONCENTRATION

Simple Models of Plant Response to Carbon Dioxide Concentration

Our primary interest in plant response to [CO_2] is to predict how plants will respond to future global climate change including the increasing global [CO_2], which is the principle driver of that change. It may not be necessary to know the detailed physiological response of plants to [CO_2] if it can be predicted from a simple equation. After examining 770 observations on the response of plant biomass to [CO_2], Kimball (1983a, b) concluded that a doubling of [CO_2] from 330 to 660 μmol mol^{-1} would probably increase agricultural yield by about 32%. In a similar exercise, Allen et al. (1987) pooled data on soybean photosynthetic rate, biomass, and seed yield from various locations. They found that the pooled normalized data for each of these variables plotted against [CO_2] could be fitted with a rectangular hyperbola, which suggested that a doubling of CO_2 would increase soybean yield about 30%, close to the estimate of Kimball. Cure (1985) also reviewed the available literature on plant response to CO_2 for 10 crops. Her study showed biomass increases of about 37% for C_3 plants and 9% for C_4 plants with a doubling of CO_2.

The papers by Kimball (1983a, b), Cure (1985), and Allen et al. (1987) suggest that plant response can be described as linear, quadratic, or hyperbolic functions of [CO_2]. These, however, are oversimplifications, because most CO_2 experiments are performed with adequate water and nutrients in contrast with the real world where these factors are generally limiting. This is especially true of natural ecosystems and of agricultural systems in poorer countries. However, even in industrialized countries, farmers will probably need to apply more fertilizer to take full advantage of the higher [CO_2]. If the climate does change as predicted (Taylor & MacCracken, Chapter 1, this book), then other environmental factors may become limiting. For example, very high temperatures could inhibit the biochemistry of growth processes, just as low temperatures do. It is, therefore, unwise to use these simple models to derive a CO_2 response factor and to then multiply measured or predicted yields by that factor. This is done frequently (e.g., Waggoner, 1984), but the results can be misleading.

Interactions between Carbon Dioxide Concentration and Other Variables

Cure (1985) and Cure and Acock (1986) examined the interactions between [CO_2] and other factors, which are discussed further by Idso (Chapter 5, this book) and by Goudriaan and Unsworth (Chapter 8, this book). Cure's literature survey revealed that CO_2 interaction data were sparse. However, the general picture that emerged was that plant response to high [CO_2] is most severely limited by nutrient stress, followed by light and temperature. There is a three-way interaction between light, temperature, and [CO_2]: the limiting effect of temperature on [CO_2] response is greater in high light than in low light. Water stress decreases crop growth and yield in all [CO_2], but, in the few studies of this interaction, plants in high [CO_2] were less inhibited by water stress than those in ambient [CO_2].

Because these interactions are complex and the existing database is so meager, we need to conduct more research to understand the mechanisms of plant responses to [CO_2] and other environmental factors. The task is daunting. After years of agricultural research and millions of field experiments, we still cannot quantify completely the response of plants to seasonal variations in light, temperature, water, and nutrients, and to the interactions between them. Now we are adding [CO_2], a variable that is expensive to control and normally involves the use of enclosures. There is no way that we can observe all possible interactions for all crop plants. Our only chance of predicting correctly what will happen to plant growth and development in future climates is to build crop simulators or models that incorporate our understanding of how each environmental factor affects the essential processes in the plant.

Of course, that understanding is presently imperfect, so current simulators have some empirical patches in them. This empiricism limits their predictive ability. To improve this situation we need to concentrate our resources on investigating the processes at work in plants and their environment, rather than continuing to demonstrate that extreme treatments cause significant, though unexplained, differences in plant response. Apart from the development of crop simulators, no other method of predicting crop growth and yield in future climates is even feasible.

Mechanistic Crop Simulators

Good progress is being made in the development of these mechanistic crop simulators (e.g., Acock et al., 1985; Wilkerson et al., 1983). For several crops, there are models that respond realistically to light, temperature, wind, humidity, soil water, and soil N. Phenology is still handled rather empirically, and, as discussed above, we have little idea what processes control C partitioning. The processes controlling branching and organ abscission are also a mystery. As for the response to [CO_2], models of leaf photosynthesis that respond to [CO_2] abound, but few crop photosynthesis models have that ability yet. We still do not understand stomatal control well enough to predict the effect of [CO_2] on stomatal resistance. Likewise, we know very little about how tissues adapt over the long term to being grown in high [CO_2]. Although we have learned a lot about the long-term response of crops to [CO_2] recently, and we are actively developing mechanistic crop simulators, we know that those simulators still have significant flaws.

We also know that the general circulation models (GCMs) used to make predictions about future climate change are flawed. For instance, the grid on GCMs is too coarse to allow clouds to be modeled mechanistically (Grotch, 1988). For several reasons, therefore, we are not yet in a position to make impact assessments that have any real meaning. Parry and his collaborators in the International Institute for Applied Systems Analysis and the United Nations Environment Program have reached the same conclusion (Parry et al., 1988). Any results obtained by exercising climate and crop simulators in their present state of development are probably misleading. Rather than looking at the results, we should be looking at the mechanisms and hypotheses incorporated in the simulators.

CONCLUSIONS

Few scientists are involved in developing crop simulators at present (1990). There is a need for scientists who specialize in various parts of the plant, soil, and atmosphere, to check that existing models accurately describe the processes and to add or amend as necessary. For that to happen, modelers must make their models more accessible to other scientists by documenting them completely, grouping the processes in a logical modular structure, and putting adequate comments in the code. Experimenters must try addressing areas of ignorance rather than further refining our knowledge of familiar phenomena. There should be more emphasis on understanding processes and less on demonstrating significant differences between treatments. This does not mean that everyone should become a biochemist; there is much biochemistry reported today that is just as empirical as most field experiments. The vital question for deciding the utility of an experiment should be, "Will it help us to predict. . .?" Ultimately, that is what agricultural research is all about.

The U.S. Department of Energy in collaboration with the USDA has been pursuing this approach with the limited resources at its disposal. It is to its credit that the DOE resisted the pressure to come up with an impact assessment before crop models with appropriate responses to [CO_2] were available. However, assessments are needed to guide policy decisions. The agricultural research community has the resources to quickly refine crop models to the point where they can be used with confidence to predict crop response to climate change, if only we can harness those resources.

REFERENCES

Acock, B., and L.H. Allen, Jr. 1985. Crop responses to elevated carbon dioxide concentrations. p. 54–97. *In* B.R. Strain and J.D. Cure (ed.) Direct effects of increasing carbon dioxide on vegetation. DOE/ER-0238. U.S. Dep. of Energy, Washington, DC. Available from NTIS, Springfield, VA.

Acock, B., and D. Pasternak. 1986. Effects of CO_2 concentration on composition, anatomy and morphology of plants. p. 41–52. *In* H.Z. Enoch and B.A. Kimball (ed.) Carbon dioxide enrichment of greenhouse crops. Vol. 2. CRC Press, Boca Raton, FL.

Acock, B., V.R. Reddy, F.D. Whisler, D.N. Baker, J.M. McKinion, H.F. Hodges, and K.J. Boote. 1985. The soybean crop simulator GLYCIM: Model documentation 1982. PB 85171163/AS. U.S. Dep. of Energy and USDA, Washington, DC. Available from NTIS, Springfield, VA.

Akita, S., and D.N. Moss. 1973. Photosynthetic responses to CO_2 and light by maize and wheat leaves adjusted for constant stomatal apertures. Crop Sci. 13:234–237.

Allen, L.H., Jr., K.J. Boote, J.W. Jones, P.H. Jones, R.R. Valle, B. Acock, H.H. Rogers, and R.C. Dahlman. 1987. Response of vegetation to rising carbon dioxide: Photosynthesis, biomass, and seed yield of soybean. Global Biogeochem. Cycles 1:1–14.

Beer, S. 1986. The fixation of inorganic carbon in plant cells. p. 3–11. *In* H.Z. Enoch and B.A. Kimball (ed.) Carbon dioxide enrichment of greenhouse crops. Vol. 2. CRC Press, Boca Raton, FL.

Cave, G., L.C. Tolley, and B.R. Strain. 1981. Effect of carbon dioxide enrichment on chlorophyll content, starch content and starch grain structure in *Trifolium subterraneum* leaves. Phys. Plant. 51:171–174.

Cure, J.D. 1985. Carbon dioxide doubling responses: A crop survey. p. 100–116. *In* B.R. Strain and J.D. Cure (ed.) Direct effects of increasing carbon dioxide on vegetation. DOE/ER0238. U.S. Dep. of Energy, Washington, DC. Available from NTIS, Springfield, VA.

Cure, J.D., and B. Acock. 1986. Crop responses to carbon dioxide doubling: A literature survey. Agric. For. Meteorol. 38:127–145.

Del Castillo, D., B. Acock, V.R. Reddy, and M.C. Acock. 1989. Elongation and branching of roots on soybean plants in a carbon dioxide-enriched aerial environment. Agron. J. 81:692–695.

Enoch, H.Z., and B.A. Kimball. 1986. Carbon dioxide enrichment of greenhouse crops. Vol. 2. Physiology, yield and economics. CRC Press, Boca Raton, FL.

Farquhar, G.D., and T.D. Sharkey. 1982. Stomatal conductance and photosynthesis. Ann. Rev. Plant Physiol. 33:317–345.

Grotch, S.L. 1988. Regional intercomparisons of general circulation model predictions and historical climate data. DOE/NBB-0084. U.S. Dep. of Energy, Washington, DC. Available from NTIS, Springfield, VA.

Hesketh, J.D., E.M. Larson, A.J. Gordon, and D.B. Peters. 1983. Internal factors influencing photosynthesis and respiration. p. 381–411. *In* J.E. Dale and F.L. Milthorpe (ed.) The growth and functioning of leaves. Cambridge Univ. Press, New York.

Ho, L. 1978. The regulation of carbon transport and the carbon balance of mature tomato leaves. Ann. Bot. 42:155–164.

Kimball, B.A. 1983a. Carbon dioxide and agricultural yield: An assemblage and analysis of 430 prior observations. Agron. J. 75:779–788.

Kimball, B.A. 1983b. Carbon dioxide and agricultural yield: An assemblage and analysis of 770 prior observations. WCL Report 14. U.S. Water Conserv. Lab., Phoenix, AZ.

Lemon, E.R. (ed.). 1983. CO_2 and plants: The response of plants to rising levels of atmospheric carbon dioxide. AAAS Selected Symp. Ser. 84. Westview Press, Boulder, CO.

Parry, M.L., T.R. Carter, and N.T. Konijn (ed.). 1988. The impact of climatic variations on agriculture. Vol. 1. Assessments in cool temperate and cold regions. Kluwer Academic Publishers, Hingham, MA.

Rogers, H.H., G.E. Bingham, J.D. Cure, J.M. Smith, and K.A. Surano. 1983. Responses of selected plant species to elevated carbon dioxide in the field. J. Environ. Qual. 12:569–574.

Strain, B.R., and J.D. Cure (ed.). 1985. Direct effects of increasing carbon dioxide on vegetation. DOE/ER-0238. U.S. Dep. of Energy, Washington, DC. Available from NTIS, Springfield, VA.

Szaniawski, R.K. 1983. Adaptation and functional balance between shoot and root activity in sunflower plants grown at different root temperatures. Ann. Bot. 51:453–459.

Thomas, J.F., and C.N. Harvey. 1983. Leaf anatomy of four species grown under continuous CO_2 enrichment. Bot. Gaz. 144:303–309.

Thornley, J.H.M. 1972. A balanced quantitative model for root:shoot ratios in vegetative plants. Ann. Bot. 36:431–441.

Tolbert, N.E., and I. Zelitch. 1983. Carbon metabolism. p. 21–64. *In* E.R. Lemon (ed.) CO_2 and plants: The response of plants to rising levels of atmospheric carbon dioxide. AAAS Selected Symp. Ser. 84. Westview Press, Boulder, CO.

Waggoner, P.E. 1984. Agriculture and carbon dioxide. Am. Sci. 72:179–184.

Wilkerson, G.G., J.W. Jones, K.J. Boote, K.T. Ingram, and J.W. Mishoe. 1983. Modeling soybean growth for crop management. Trans. ASAE 26:63–73.

5 Interactive Effects of Carbon Dioxide and Climate Variables on Plant Growth[1]

Sherwood B. Idso

U.S. Water Conservation Laboratory
Phoenix, Arizona

ABSTRACT

The climate variables predicted to experience major modification as a result of future increases in atmospheric CO_2 and other radiatively active trace gases are temperature and precipitation. Predicted changes in these two parameters should intensify the hydrologic cycle over the globe, but could produce opposite trends in certain regions. A "worst-case" scenario of consequent local reductions in summer soil moisture is evaluated in terms of the beneficial effects of atmospheric CO_2 enrichment on plant water use efficiency and the interactive effect of air temperature increase on the growth-enhancing effects of atmospheric CO_2 enrichment. It is demonstrated that the direct biological impacts of concomitant increases in CO_2 and air temperature are probably sufficient to offset the adverse effects of summer soil moisture reductions predicted by state-of-the-art climate/water balance models. It is also noted that the worst-case climate scenario is unrealistic. Consequently, plant growth the world over should be significantly stimulated by atmospheric CO_2 enrichment, a phenomenon that many people feel is already evident in a number of ecological indicators.

In the absence of concomitant climatic changes, projected increases in the CO_2 content of the Earth's atmosphere are generally acknowledged to be of a positive nature in terms of promoting plant growth in general and agricultural yield in particular (Wittwer & Robb, 1964; Lemon, 1983; Strain & Cure, 1985; Acock, Chapter 4, this book; Patterson & Flint, Chapter 7, this book; Goudriaan & Unsworth, Chapter 8, this book). But what happens if the climate changes? Two major problems immediately come to mind. First of all, there is the basic worry that higher air temperatures resulting from the CO_2/trace gas "greenhouse effect" may have a negative impact on plant

[1] Contribution from the USDA-ARS.

 Impact of Carbon Dioxide, Trace Gases, and Climate Change on Global Agriculture, ASA Special Publication no. 53.

growth. And secondly, although the hydrologic cycle is expected to intensify on a global basis (Schlesinger & Mitchell, 1985), localized reductions in precipitation have been predicted. Indeed, several climate models predict significant decreases in summer rainfall and, hence, summer soil moisture, over extensive regions of middle and high latitudes in the Northern Hemisphere, such as the North American Great Plains, western Europe, northern Canada, and Siberia (Manabe et al., 1981; Mitchell & Lupton, 1984; Manabe & Wetherald, 1987). Other climate models, however, predict just the opposite, with increased summer soil moisture over much of North America, Asia, and Europe (Hansen et al., 1984; Washington & Meehl, 1984; Schlesinger & Mitchell, 1987). Nevertheless, it is important to consider the ramifications of the negative projections, because several studies have indicated that changes in soil water content are probably of even more importance to agriculture than are changes in surface air temperature (Manabe & Wetherald, 1986; Gleick, 1987a; Bultot et al., 1988).

SOIL MOISTURE DEPLETION

So what is the worst-case scenario for decreased summer soil moisture? Before we can answer this question, we need to know exactly what is meant by the term *soil moisture* in studies of this phenomenon. Probably the clearest definition is that given by Gleick (1987b): "The standard soil moisture term in water balances is 'available water capacity', typically the maximum amount of water that can be stored in the soil and removed by evaporation or vegetation demands," that is, "the water that. . .soils can hold and make available for crops and other flora."

On this basis, Gleick (1987a) calculated that the plant-available soil moisture in the agricultural portion of the Sacramento Basin of northern California could be decreased by 8 to 44% for a 300 to 600 μmol CO_2 per mol air doubling of the atmospheric CO_2 concentration. In the global studies of Manabe and Wetherald (1986, 1987), however, maximum soil moisture decreases as great as 50% were obtained for certain regions of the world as a result of such a rise in the air's CO_2 content.

PLANT WATER USE EFFICIENCY

With no changes in other factors, worst-case reductions in soil water content would be disastrous, as most of the climate modelers who have predicted such decreases have intimated. It is a well-established fact, however, that coincident with such a reduction in soil moisture, there would be an increase in the per-unit-leaf-area water use efficiency of practically all agricultural crops (Acock & Allen, 1985). The mean increase in this parameter has been documented to be an approximate doubling for a doubling of the atmospheric CO_2 content, all else being equal (Rogers et al., 1983; Kimball & Idso, 1983). Consequently, with plant leaves able to use soil moisture twice as efficiently at an atmospheric CO_2 concentration of 600 μmol CO_2 mol^{-1} air as they do at a concentration of 300 μmol CO_2 mol^{-1} air, a 50% reduction in *plant-available* soil moisture would have no net effect on agricultural productivity if plant leaf area did not increase as it normally does with CO_2 enrichment under well-watered conditions. Hence, the doubled water use efficiency of such an unchanged leaf area should be able to compensate for the "worst-case" 50% reduction in available soil water in this "best-case" plant response scenario. Accordingly, more realistic (less dramatic) reductions in soil water content are likely to be adequately countered by the more realistic (less dramatic) increases in plant water use efficiency, which occur when this parameter is evaluated on a per-unit-land-area basis, which allows for an increase in leaf area under conditions of CO_2 enrichment.

Recently, however, it has been suggested that by the time the CO_2 concentration of the atmosphere reaches 600 μmol CO_2 mol^{-1} air, the combined climatic effect of other concomitantly increasing trace gases will be equivalent to that of CO_2 (Ramanathan et al., 1985, 1987; Taylor & MacCracken, Chapter 1, this book; Burke & Lashof, Chapter 3, this book). This phenomenon would again lead to disastrous consequences for agriculture in the worst-case scenario, because these trace gases do not have the positive effect on plant water use efficiency, which CO_2 has. Nevertheless, there are other factors that may offset the deleterious effects of additional trace gas-induced soil moisture deficiencies. In fact, the increasing air temperature may provide a key element of the needed relief.

POSITIVE ASPECTS OF GLOBAL WARMING

Consider the inhibitory effect of normal oxygen concentrations on CO_2 uptake by plants (Forrester et al., 1966; Hesketh, 1967; Bowes & Ogren, 1972; Ehleringer, 1979; Acock, Chapter 4, this book). A number of different experiments have established that this O_2-inhibition of photosynthesis can be almost totally eliminated by raising the CO_2 content of the atmosphere to 900 μmol CO_2 mol^{-1} air (Jensen, 1977; Mortensen, 1983; Mortensen & Moe, 1983; Mortensen & Ulsaker, 1985). Thus, because the O_2 inhibition of photosynthesis increases with increasing air temperature (Joliffe & Tregunna, 1968; Laing et al., 1974; Monson et al., 1982), and because it can be almost totally eliminated by atmospheric CO_2 enrichment, it follows that there is a greater potential for benefit at higher air temperatures. In the words of Mortensen (1987), "the [positive] effect of CO_2 enrichment will increase with increasing temperatures," and "the optimum temperature for photosynthesis will, as a consequence, increase in CO_2-enriched air."

Figure 5-1 provides a manifestation of this phenomenon in two agricultural crops [bell pepper (*Capsicum annuum* L. var. *annuum*) and tomato (*Lycopersicon esculentum* Mill.)], plus a cottonwood tree (*Populus fremontii*), as observed by Sage and Sharkey (1987). As is readily evident, at a temperature of 10 °C there is little change in CO_2 assimilation rate as the intercellular CO_2 content is increased from 300 to 600 μmol CO_2 mol^{-1} air.

At 20 °C, however, such a CO_2 doubling has a significant positive impact on net photosynthesis, and at 30 °C its positive effects are greatest of all.

To more readily appreciate this fact, I have recast the data of Fig. 5-1 in the format of Fig. 5-2, where I have plotted the ratio of net photosynthesis at 600 μmol CO_2 mol^{-1} air to net photosynthesis at 300 μmol CO_2 mol^{-1} air as a function of leaf temperature. And as this figure indicates, although atmospheric CO_2 enrichment has little or no impact on net photosynthesis at 10 °C, it has a significant stimulatory effect, which becomes greater and greater as leaf temperature rises above that value.

My associates and I have also observed this interactive phenomenon with a number of plants maintained out-of-doors at Phoenix, AZ, in open-top, clear-plastic-wall, CO_2-enrichment chambers (Idso et al., 1987). Under such naturally varying environmental conditions, however, it is difficult to separate the effects of air temperature from those of solar radiation. Thus, we have sometimes used the mean net photosynthetic rate of our different CO_2 treat-

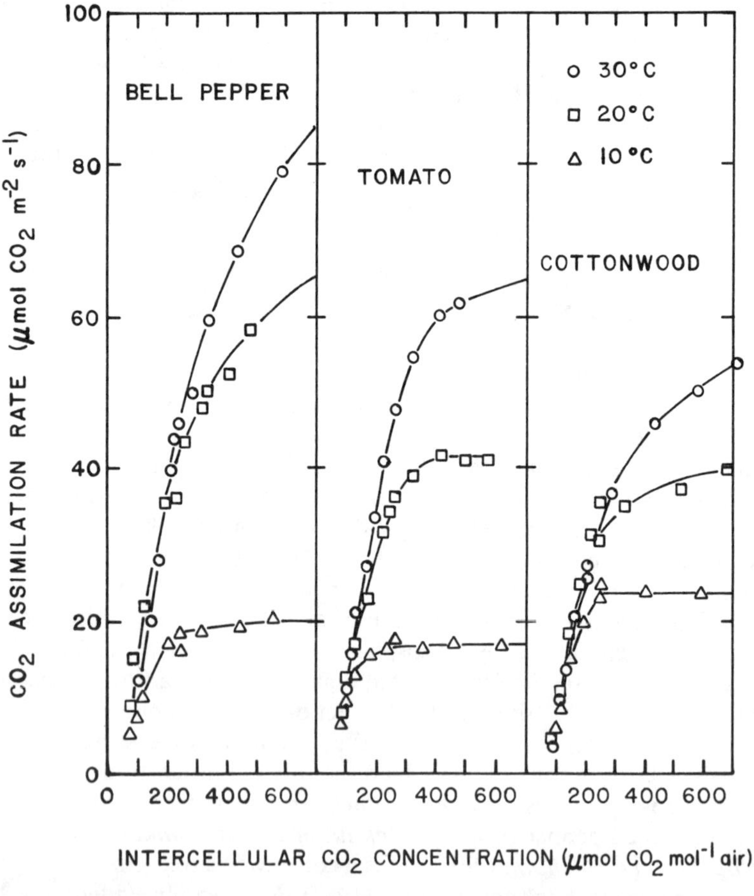

Fig. 5-1. Net photosynthesis (CO_2 assimilation rate) as a function of intercellular CO_2 concentration at three different leaf temperatures for bell pepper, tomato, and cottonwood tree. Adapted from Sage and Sharkey (1987).

ments as a surrogate for the combined influence of all environmental parameters that may impact plant growth. By plotting individual treatment net photosynthesis rates as functions of their combined mean rate, as in Fig. 5-3, for example, the short-term effects observed with waterlily (*Nymphea marliac carnea*) leaves over several diurnal cycles are seen to be of the same basic nature as the long-term effects observed with waterfern (*Azolla pinnata pinnata*) plants over the greater part of a year. That is, when conditions are not conducive to high growth rates, atmospheric CO_2 enrichment has little effect on net photosynthesis, as was also evident in the data of Sage and Sharkey (1987). But as conditions improve, as manifest by the increased photosynthetic rates of all treatments, the stimulatory effects of added CO_2 become ever greater.

In one of our first studies of this nature, we obtained essentially identical results under both sunlit and shaded conditions, indicative of the likelihood that the air temperature variability over the year was the primary factor responsible for the type of response evident in Fig. 5–3 (Idso et al., 1987). Hence, we have presented our newest results for carrot (*Daucus carota* L.) and radish (*Raphanus sativus* L.) as shown in Fig. 5–4, where the growth modification factor is the weekly dry weight production at 640 μmol CO_2

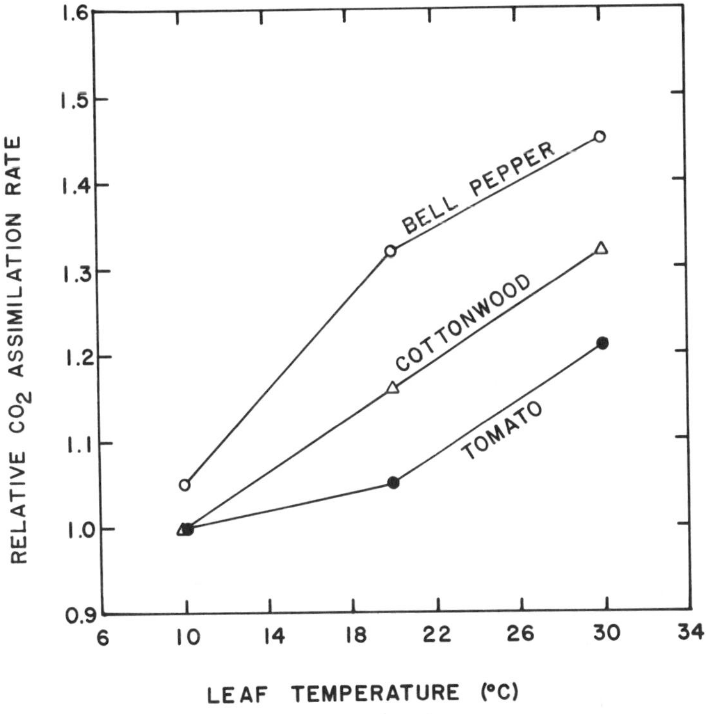

Fig. 5–2. Relative CO_2 assimilation rate (net photosynthesis at an intercellular CO_2 concentration of 600 μmol CO_2 mol^{-1} air divided by net photosynthesis at 300 μmol CO_2 mol^{-1} air) as a function of leaf temperature for bell pepper, tomato, and cottonwood tree as determined from the data of Fig. 5–1.

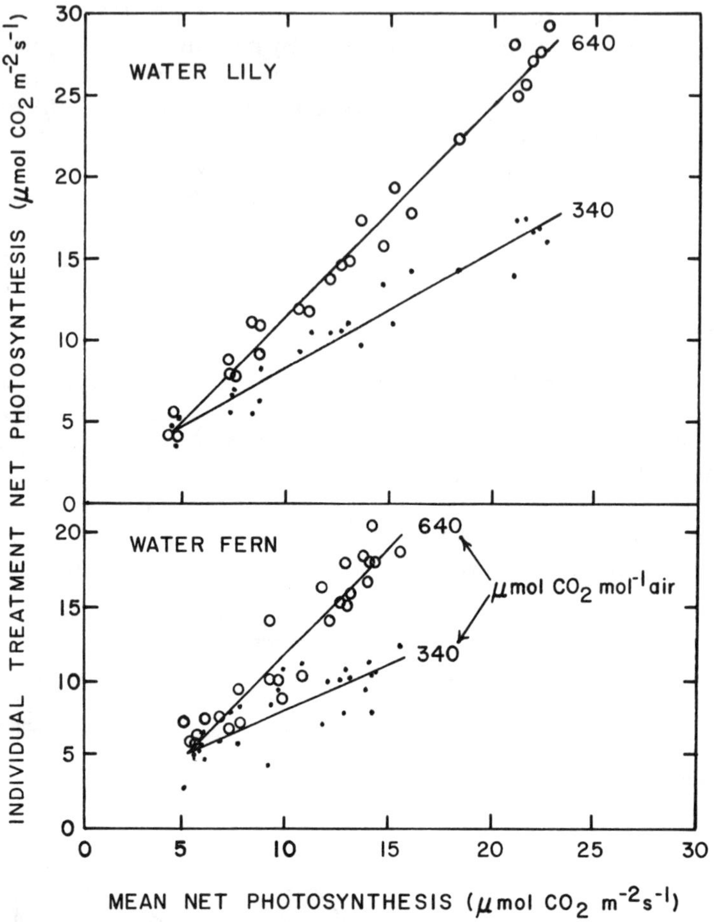

Fig. 5–3. Regressions of individual treatment net photosynthesis rates on the mean net photosynthetic rate of both treatments (340 and 640 μmol CO_2 mol^{-1} air) for waterlily leaves sampled over a several-day period (top) and waterfern plants sampled over a several-month period (bottom). Adapted from Allen et al. (1988a, b).

mol^{-1} air divided by the weekly dry weight production at 340 μmol CO_2 air, averaged over a large number of individual experiments. As this figure demonstrates, the CO_2-temperature interaction is again like that of the preceding studies. The effects of added CO_2 are negligible in the vicinity of 10 °C, but they become greater and greater as temperature rises from that point. When the optimum temperature for plant growth is exceeded (which occurred at about 30 °C for carrot and at about 26 °C for radish in this experiment), the *relative* plant growth enhancement due to atmospheric CO_2 enrichment continues to increase with increasing air temperature. The absolute growth rate, however, may be reduced. In addition, it is clear from this figure that there may be significant differences among species with respect to the rate at which the stimulatory effect of atmospheric CO_2 enrichment increases with increasing air temperature.

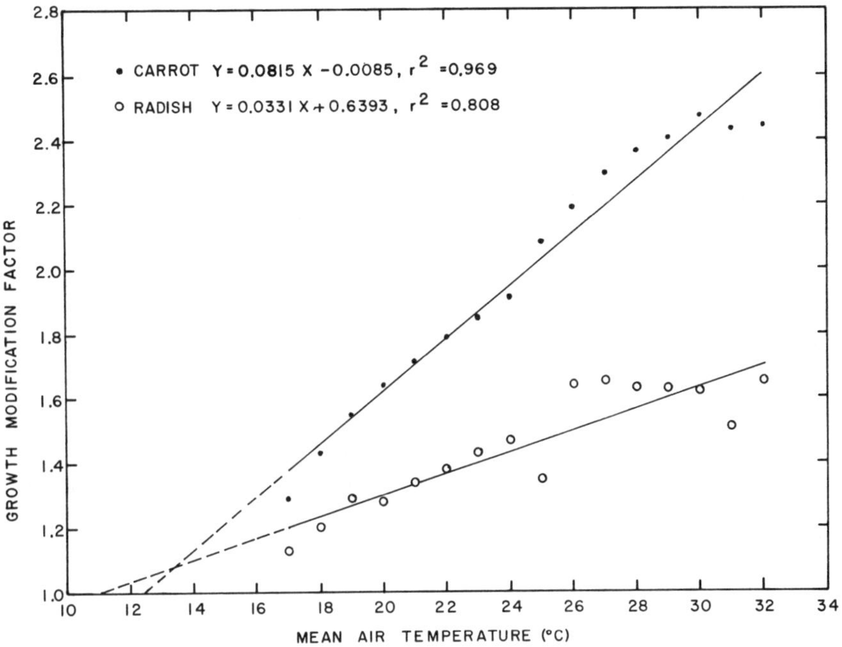

Fig. 5–4. Growth modification factors (weekly dry wt. production at 640 μmol CO_2 mol^{-1} air divided by weekly dry wt. production at 340 μmol CO_2 mol^{-1} air) for carrot and radish plants as functions of the mean air temperature for a number of experiments conducted over the year at Phoenix, AZ. Adapted from Idso and Kimball (1989).

CONCLUSIONS

What are the ramifications of this effect? Consider the results of Fig. 5–4. The results were derived from actual plant growth data, which is the bottom-line parameter of most significance for agriculture. First, locate the temperature at which each of the two plants there depicted have growth modification factors of 1.3—the standard mean response for a 300 μmol CO_2 mol^{-1} air increase in the CO_2 content of the atmosphere (Kimball, 1983a, b). Then follow the two relationships along for another 4 °C—the mean global warming generally predicted for such a CO_2 increase (Schlesinger & Mitchell, 1987)—and a new *mean* growth modification factor of 1.53 is obtained. Continuing for another 4 °C, to accommodate the additional trace gas warming, which is predicted to accompany that due to CO_2 alone, a final "best-case" mean growth modification factor of 1.76 is obtained. And because the 76% enhancement of plant growth determined from this analysis is more than 2.5-fold greater than that expected under unchanging climatic conditions, it should be more than sufficient to compensate for the extra worst-case soil moisture depletions predicted to result from non-CO_2 trace gas warming. It is possible, however, that the deleterious direct effects of very high temperatures could largely cancel the positive effects of this interaction in terms of absolute plant growth rates.

But it must be reiterated that these worst-case climate changes may never actually occur, as previously discussed (Idso, Chapter 2, this book). Hence, it is likely that the flora of the Earth will be significantly benefited by a continuation of the current upward trend in atmospheric CO_2 content. Indeed, although too voluminous to treat in this brief review, there is much evidence to suggest that a dramatic CO_2-induced stimulation of the biosphere is already in progress (Gifford, 1979; Wittwer, 1982; Lamarche et al., 1984; Idso, 1986, 1989; Allen et al., 1987; Morison, 1987; Woodward, 1987, 1988).

REFERENCES

Acock, B., and L.H. Allen, Jr. 1985. Crop responses to elevated carbon dioxide concentrations. p. 53–97. *In* B.R. Strain and J.D. Cure (ed.) Direct effects of increasing carbon dioxide on vegetation. DOE/ER-0238. U.S. Dep. of Energy, Washington, DC.

Allen, L.H., Jr., K.J. Boote, J.W. Jones, P.H. Jones, R.R. Valle, B. Acock, H.H. Rogers, and R.C. Dahlman. 1987. Response of vegetation to rising carbon dioxide: Photosynthesis, biomass, and seed yield of soybean. Global Biogeochem. Cycles 1:1–14.

Allen, S.G., S.G. Idso, and B.A. Kimball. 1988a. Photosynthetic response of two aquatic plant species to elevated carbon dioxide. J. Ariz. Nev. Acad. Sci. Proc. Suppl. 23:8.

Allen, S.G., S.B. Idso, B.A. Kimball, and M.G. Anderson. 1988b. Interactive effects of CO_2 and environment on photosynthesis of *Azolla*. Agric. For. Meteorol. 42:209–217.

Bowes, G., and W.L. Ogren. 1972. Oxygen inhibition and other properties of soybean ribulose-1,5-biphosphate carboxylase. J. Biol. Chem. 247:2171.

Bultot, F., G.L. Dupriez, and D. Gellens. 1988. Estimated annual regime of energy-balance components, evapotranspiration and soil moisture for a drainage basin in the case of a CO_2 doubling. Clim. Change 12:39–56.

Ehleringer, J.R. 1979. Photosynthesis and photorespiration. Biochemistry, physiology, and ecological implications. HortScience 14:217–222.

Forrester, M.L., G. Krotkow, and C.D. Nelson. 1966. Effect of oxygen on photosynthesis, photorespiration and respiration in detached leaves. I. Soybean. Plant Physiol. 41:422–427.

Gifford, R.M. 1979. Growth and yield of CO_2-enriched wheat under water-limited conditions. Aust. J. Plant Physiol. 6:367–378.

Gleick, P.H. 1987a. Regional hydrologic consequences of increases in atmospheric CO_2 and other trace gases. Clim. Change 10:137–161.

Gleick, P.H. 1987b. The development and testing of a water balance model for climate impact assessment: Modeling the Sacramento Basin. Water Resour. Res. 23:1049–1061.

Hansen, J., L. Lacis, D. Rind, G. Russel, P. Stone, I. Fung, R. Ruedy, and J. Lerner. 1984. Climate sensitivity: Analysis of feedback mechanisms. p. 103–163. *In* J.E. Hansen and T. Takahasi (ed.) Climate processes and climate sensitivity. Am. Geophys. Union, Washington, DC.

Hesketh, J. 1967. Enhancement of photosynthetic CO_2 assimilation in the absence of oxygen, as dependent upon species and temperature. Planta 76:371–374.

Idso, S.B. 1986. Industrial age leading to the greening of the Earth? Nature (London) 320:22.

Idso, S.B. 1989. Carbon dioxide and global change: Earth in transition. IBR Press, Tempe, AZ.

Idso, S.B., and B.A. Kimball. 1989. Growth response of carrot and radish to atmospheric CO_2 enrichment. Environ. Exp. Bot. 29:135–139.

Idso, S.B., B.A. Kimball, M.G. Anderson, and J.R. Mauney. 1987. Effects of atmospheric CO_2 enrichment on plant growth: The interactive role of air temperature. Agric. Ecosyst. Environ. 20:1–10.

Jensen, R.G. 1977. Ribulose-1,5-bisphophate carboxylase oxygenase. Ann. Rev. Plant Physiol. 28:379–400.

Joliffe, P.A., and E.B. Tregunna. 1968. Effect of temperature, CO_2 concentration, and light intensity on oxygen inhibition of photosynthesis in wheat leaves. Plant Physiol. 43:902–906.

Kimball, B.A. 1983a. Carbon dioxide and agricultural yield: An assemblage and analysis of 430 prior observations. Agron. J. 75:779–788.

Kimball, B.A., and S.B. Idso. 1983b. Carbon dioxide and agricultural yield: An assemblage and analysis of 770 prior observations. WCL Rep. 14. U.S. Water Conserv. Lab., Phoenix, AZ.

Kimball, B.A., and S.B. Idso. 1983. Increasing atmospheric CO_2: Effects on crop yield, water use, and climate. Agric. Water Manage. 7:55–73.

Laing, W.A., W.L. Ogren, and R.H. Hageman. 1974. Regulation of soybean net photosynthetic CO_2 fixation by the interaction of CO_2, O_2, and ribulose 1,5-diphosphate carboxylase. Plant Physiol. 54:678–685.

LaMarche, V.C., Jr., D.A. Graybill, H.C. Fritts, and M.R. Rose. 1984. Increasing atmospheric carbon dioxide: Tree-ring evidence for growth enhancement in natural vegetation. Science (Washington, DC) 225:1019–1021.

Lemon, E.R. (ed.). 1983. CO_2 and plants: The response of plants to rising levels of atmospheric carbon dioxide. Westview Press, Boulder, CO.

Manabe, S., and R.T. Wetherald. 1986. Reduction in summer soil wetness induced by an increase in atmospheric carbon dioxide. Science (Washington, DC) 232:626–628.

Manabe, S., and R.T. Wetherald. 1987. Large-scale changes of soil wetness induced by an increase in atmospheric carbon dioxide. J. Atmos. Sci. 44:1211–1235.

Manabe, S., R.T. Wetherald, and R.J. Stouffer. 1981. Summer dryness due to an increase of atmospheric CO_2 concentration. Clim. Change 3:347–386.

Mitchell, J.F.B., and G. Lupton. 1984. A 4 × CO_2 integration with prescribed changes in sea surface temperatures. Prog. Biometeorol. 3:353–374.

Monson, R.K., M.A. Stidham, G.J. Williams III, G.E. Edwards, and E.G. Uribe. 1982. Temperature dependence of photosynthesis in *Agropyron smithii* Rydb. Plant Physiol. 69:921–928.

Morison, J.I.L. 1987. Plant growth and CO_2 history. Nature (London) 327:560.

Mortensen, L.M. 1983. Growth responses of some greenhouse plants to environment. VIII. Effect of CO_2 on photosynthesis and growth of Norway spruce. Meld. Nor. Landbrukshoegsk. 62:1–13.

Mortensen, L.M. 1987. Review: CO_2 enrichment in greenhouses. Crop responses. Sci. Hortic. 33:1–25.

Mortensen, L.M., and R. Moe. 1983. Growth responses of some greenhouse plants to environment. V. Effect of CO_2, O_2 and light on net photosynthetic rate in *Chrysanthemum morifolium* Ramat. Sci. Hortic. 19:133–140.

Mortensen, L.M., and R. Ulsaker. 1985. Effect of CO_2 concentration and light levels on growth, flowering and photosynthesis of *Begonia* × *hiemalis* Fotsch. Sci. Hortic. 27:133–141.

Ramanathan, V., L. Callis, R. Cess, J. Hansen, I. Isaksen, W. Kuhn, A. Lacis, F. Luther, J. Mahlman, R. Reck, and M. Schlesinger. 1987. Climate-chemical interactions and effects of changing atmospheric trace gases. Rev. Geophys. 25:1441–1482.

Ramathan, V., R.J. Cicerone, H.B. Singh, and J.T. Kiehl. 1985. Trace gas trends and their potential role in climate change. J. Geophys. Res. 90:5547–5566.

Rogers, H.H., J.F. Thomas, and G.M. Bingham. 1983. Response of agronomic and forest species to elevated atmospheric carbon dioxide. Science (Washington, DC) 220:428–429.

Sage, R.F., and T.D. Sharkey. 1987. The effect of temperature on the occurrence of O_2 and CO_2 insensitive photosynthesis in field grown plants. Plant Physiol. 84:658–664.

Schlesinger, M.E., and J.F.B. Mitchell. 1985. Model projections of the equilibrium climatic response to increased carbon dioxide. p. 81–147. *In* M.C. MacCracken and F.M. Luther (ed.) Projecting the climatic effects of increasing carbon dioxide. U.S. Dep. of Energy, Washington, DC. Available from NTIS, Springfield, VA.

Schlesinger, M.E., and J.F.B., Mitchell. 1987. Climate model simulations of the equilibrium climatic response to increased carbon dioxide. Rev. Geophys. 25:760–798.

Strain, B.R., and J.D. Cure (ed.) 1985. Direct effects of increasing carbon dioxide on vegetation. U.S. Dep. of Energy, Washington, DC. Available from NTIS, Springfield, VA.

Washington, W.M., and G.A. Meehl. 1984. Seasonal cycle experiment on the climate sensitivity due to a doubling of CO_2 with an atmospheric general circulation model coupled to a simple mixed-layer ocean model. J. Geophys. Res. 89:9475–9503.

Wittwer, S.H. 1982. Carbon dioxide and crop productivity. New Sci. 95:233–234.

Wittwer, S.H., and W.M. Robb. 1964. Carbon dioxide enrichment of greenhouse atmospheres for food crop production. Econ. Bot. 18:34–56.

Woodward, F.I. 1987. Stomatal numbers are sensitive to increases in CO_2 from pre-industrial levels. Nature (London) 327:617–618.

Woodward, F.I. 1988. The responses of stomata to changes in atmospheric levels of CO_2. Plants Today 1:132–135.

6 Predicted Impact of Climatic Warming on Soil Properties and Use[1]

S. W. Buol, P. A. Sanchez, and S. B. Weed

North Carolina State University
Raleigh, North Carolina

J. M. Kimble

USDA-SCS
Lincoln, Nebraska

ABSTRACT

Warming of the atmosphere will increase mean annual soil temperatures. Large areas in the Northern Hemisphere could become suitable for cultivation, assuming little change in rainfall patterns. A 3 °C temperature rise is predicted to cause an overall 11% decrease in soil organic matter content to a 30-cm depth in the temperature zone, but this increase is well within the standard deviations of the data. If so, a 3 °C temperature increase in the temperate zone is predicted to increase atmospheric CO_2 content by approximately 8% over the present (1990) levels during a 50-yr period, provided no increase in biomass production occurs. Carbon/N ratios of soil organic matter are expected to narrow slightly, and total N in the soil is predicted to increase by 10% in the temperate region. However, an average increase of 568 kg ha^{-1} above and belowground organic residue return to the soil is sufficient to offset changes in soil organic C content. No major changes in fertilization practices are predicted, because modern fertilizer rates greatly exceed weathering rates in providing nutrients to plants. Increased solubility of soil minerals is expected to be slight. Redder soil colors may result from hematite formation.

The guidelines from the organizing committee were to examine alteration in soil properties that could be expected with (i) a 3 °C increase in mean annual global air temperature and (ii) a 10% increase or decrease in mean an-

[1] Paper no. 12094 of the Journal Series of the North Carolina Agricultural Research Service, Raleigh, NC 27695-7643. Supported in part by NSF grant no. BSR-8717274 "Soil Organic Matter Dynamics" to North Carolina State University.

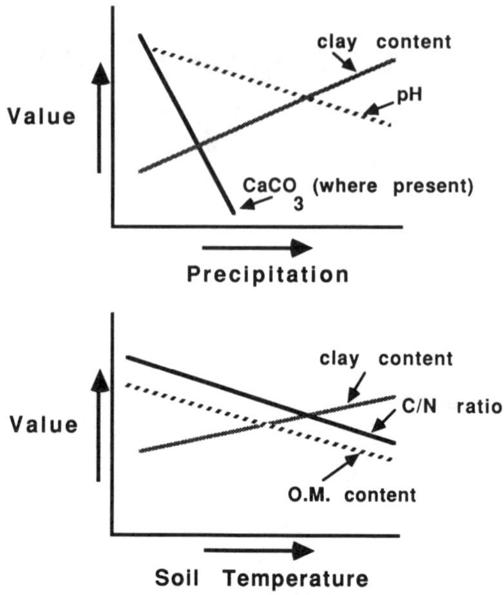

Fig. 6–1. Schematic of soil-property vs. climatic relationships.

nual rainfall. We assume that changes will be gradual and will occur over a 50-yr period.

The most likely soil property-climatic factor relationships were presented in 1941 by Jenny (1941) and are schematically represented in Fig. 6–1. Jenny expressed soil properties as functions of temperature in Eq. [1] where *S* is the soil property to be measured; *T* is the mean annual soil temperature; *m* is the precipitation and/or soil moisture; *o* is organisms (natural or cultivated); *p* is parent material; *r* is relief or geomorphic position, and *t* is time of soil formation. Following Jenny, Eq. [1] reads soil properties are a function of temperature when moisture, organisms, parent materials, relief, and time are constant.

$$S = f(T)\ m, o, p, r, t, \ldots \qquad [1]$$

METHODS

Our approach to predicting soil property change with a change in climate was to examine what is known about the fundamental chemistry of soil minerals and organic matter and how they respond to increased soil temperature, i.e., temperature of the soil water. We then used the approach that "the present is the key to the future" in which we compared soil characteristics from different soil temperature regimes (Soil Survey Staff, 1975) and assessed the impact of a 3 °C mean annual soil temperature increase. We then examined present and recent history of commercial fertilizer use and speculated on the impact soil property changes may have on new available lands and on fertilizer and liming practices.

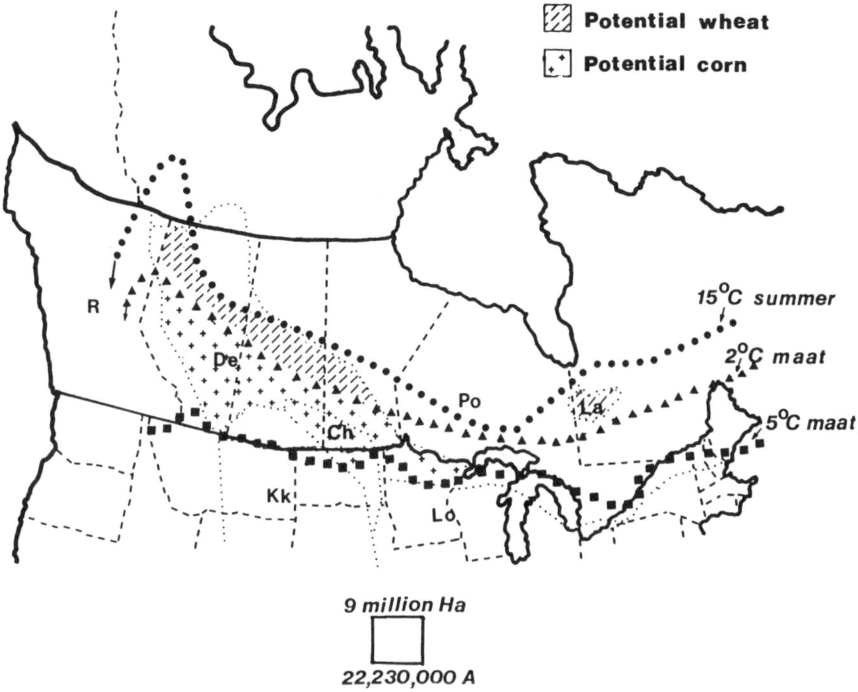

Fig. 6–2. Predicted areas of corn and wheat cropland increase in North America with a 3 °C mean annual air temperature rise. *R* and *Po*: Mostly unsuitable mountainous terrain or shallow, stony Podzols. *De, La, Lo, Ch*: High proportion suitable soils mainly Mollisols and Alfisols. *KK*: Marginal soils due to low rainfall.

Soil Temperature

Mean annual soil temperature (MAST), at any depth, is closely related but slightly higher than mean annual air temperature (MAAT) (Smith et al., 1964). For thermic and mesic soil temperature regimes (MAST 15 to 22 °C and 8 to 15 °C, respectively) in the central USA, the mean annual soil temperature is about 1.1 °C higher than mean annual air temperature. In colder regions with frigid soil temperature regimes (MAST 0 to 8 °C) the difference is greater. Although we lack a good data base, MAST appears to exceed MAAT by 3 °C. Aboveground weather parameters probably have a greater influence on farming. However, a 5 °C MAAT, which approximates the 8 °C MAST in northern latitudes, presently limits the production of corn (*Zea mays* L.) for grain.

In the absence of sufficient data about soil temperatures, we have examined the geographic change in the mesic-frigid soil family boundary, i.e., 8 °C MAST by using published MAAT (Hargreaves & Samani, 1986) and assuming 5 °C MAAT equals 8 °C MAST. We then projected climatic temperature increase of 3 °C mean MAAT. Using the 5 °C MAAT line as an approximation of the 8 °C MAST and 2 °C MAAT to approximate 5 °C MAST in the Northern Hemisphere (Fig. 6–2 and 6–3), we then overlaid this projec-

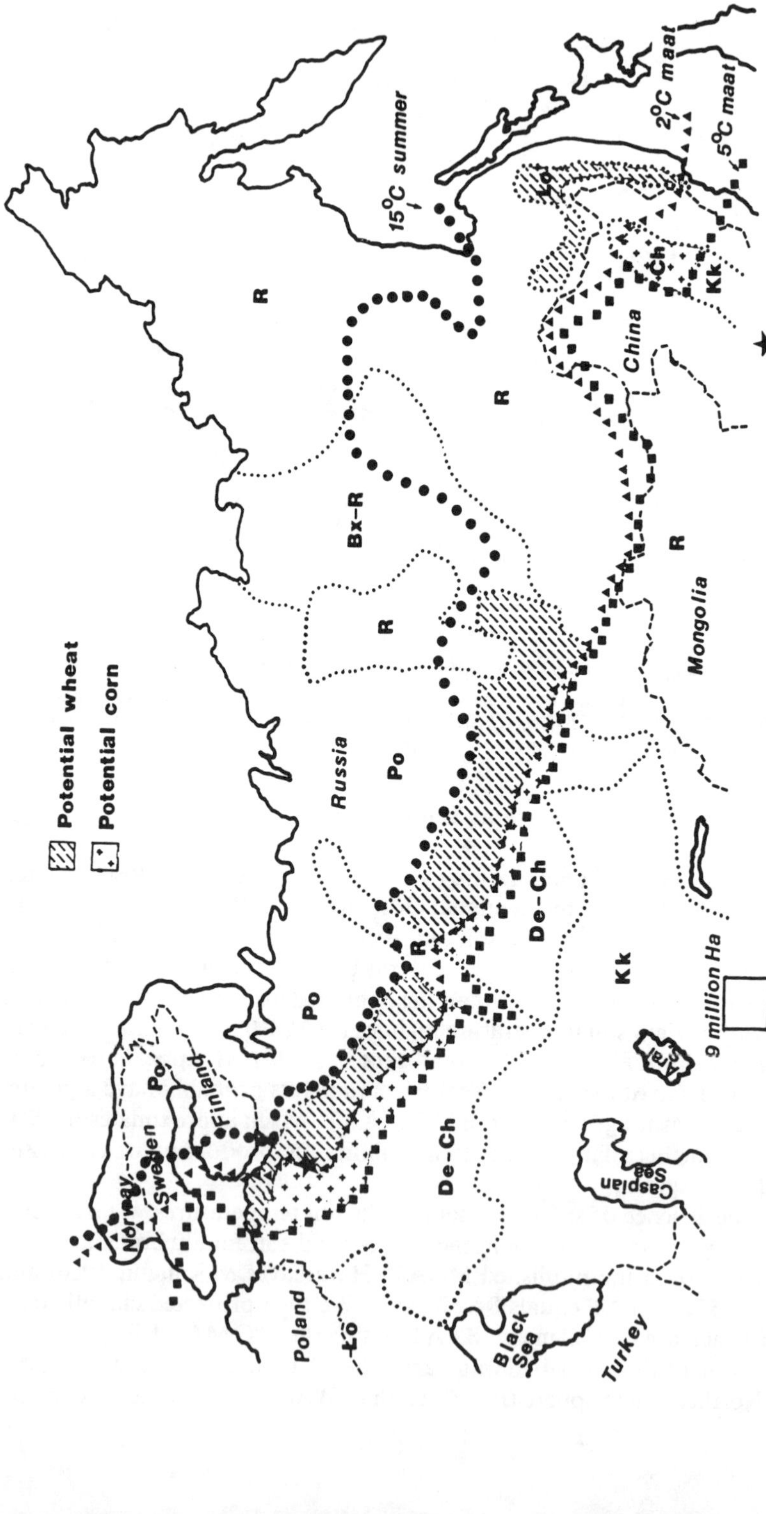

Fig. 6–3. Predicted areas of corn and wheat cropland in Europe and Asia with 3 °C rise in mean annual air temperature. *R*, *Bx-R*, *Po*: Mostly unsuitable mountainous terrain or shallow stony soils. *Ch*, *De-Ch*, *Lo*: High proportion of suitable soils, mainly Mollisols and Alfisols. *KK*: Marginal soil due to low rainfall.

Table 6–1. Potential increases in Corn Belt land with a 3 °C increase in mean annual air temperature.

Country	Current area†	Potential increase‡	New total	Increase
	million ha			%
USSR	5	30	35	600
Canada	1	31	32	3100
China	17	11	28	65
USA	30	8	38	26

† FAO, 1985.
‡ Assuming 50% of new potentially arable land is locally suitable for cultivation.

Table 6–2. Potential increases in wheat belt land with a 3 °C increase in mean annual air temperature.

Country	Current area†	Potential increase‡	New total	Increase
	million ha			%
USSR	50	71	121	142
Canada	13	19	32	146
China	29	0	29	0
USA	26	0	26	0

† FAO, 1985.
‡ Assuming 50% of new potentially arable land is locally suitable for cultivation.

tion on the FAO world soil map (FAO, 1975, 1978a, b) and outlined areas dominated by terrain and soil type desirable for commercial farming (Fig. 6–2 and 6–3). This predicted geographic northward shift can, with great caution, be considered an extension of the present corn belt with climatic warming of 3 °C. The magnitude of this area was reduced by 50% as an estimate of areas not suitable for crop production. The results compared with present production areas are presented by country in Table 6–1.

With similar caution, we have extrapolated the present wheat (*Triticum aestivum* L.)-growing areas northward, on favorable landscapes and soil types, to reflect higher summer (June, July, and August) temperatures. Present wheat-growing areas have mean air temperatures for the months of June, July, and August >18 °C. Thus, allowing for a predicted 3 °C temperature increase in each of these months, we produced the 15 °C summer line on Fig. 6–2 and 6–3. We again overlaid this line on the FAO world soil map and outlined areas dominated by suitable soils and terrain to obtain the location (Fig. 6–2 and 6–3) and extent (Table 6–2) of new wheat areas, assuming that only 50% would be suitable for production.

Although our intent is not to examine the influence of climatic warming on crop production, which is discussed by Goundriaan and Unsworth (Chapter 8, this book), we simply are drawing attention to the close relationship of mean annual soil temperature to mean annual air temperature. We then use a geographic display of cold temperature–limited farming systems to illustrate where the most significant soil use patterns may be altered with climatic warming in the northern latitudes. Any possible reduction of the corn- or wheat-growing areas because of increased temperature was not included. Such calculation is not relevant to corn as it grows well in the tropics, but may be important to wheat because it is a heat-sensitive species.

Soil Chemical and Mineralogical Properties

Soil properties that might be affected by an increase in soil temperature include solution chemical reaction rates, diffusion-controlled reactions, solubilities of solid and gaseous phases, and equilibrium states. The combined effects due to a 3 °C increase in temperature may be too small to detect in a complex soil systsem, at least on a short-term basis, although they are readily seen in pure solution. Changes induced in soil properties may be apparent only after many years under the changed environmental conditions.

Reaction rates and diffusion-controlled reactions generally increase as temperature of the solution increases. Solubilities may increase or decrease, depending on the specific material being considered. For example, the dissolution of calcite in pure water decreases with an increase in temperature, as shown by the pKsp (−log of the solubility product) = 8.35 at 5 °C; 8.53 at 40 °C). The increase in pKsp corresponds to a decrease in solubility (Stumm & Morgan, 1981). Similarly, the common gases (CO_2, N_2, O_2) all show a decrease in solubility as the temperature increases. On the other hand, the solubility of K and Na salts increases with temperature. Similar effects are undoubtedly experienced by the very slightly soluble minerals present in soils, but the magnitude of the effect is very small for the 3 °C rise considered.

The relationship between goethite and hematite illustrates an apparent effect of temperature on equilibrium states. As the temperature increases, other factors being equal, the ratio of hematite to goethite in the soil increases, and the result is a redder soil color (Kampf & Schwertmann, 1982; Schwertmann et al., 1982, Bigham et al., 1978). The long-term effect of this, pedogenically, would be to move red-colored hematitic soils further poleward in the temperate zones. Again, however, this effect would be seen only after a considerable period of pedogenic time.

The general effect of an increase in temperature on soil solution components may relate to the effect on the properties of water itself (Garrels & Christ, 1965). Increasing water temperature results in decreased density and decreased tendency to hydrate ions in solution, which results in greater ionic interaction, e.g., ion pairing. Thus, ion speciation should change. Evaluation of this effect demands calculation of the thermodynamic parameters with a new temperature base. Parameters used in programs such as GEOCHEM (Sposito & Mattigod, 1979) are standardized at 25 °C. Uncertainties in experimental values of such parameters make evaluation of the effects of a 3 °C temperature rise questionable.

Soil Organic Matter Content

Soil is a temporary repository for organic C. The organic C content of any soil reflects a steady state between accumulation of C from plant residue and oxidation of the soil organic C. Organic matter contents are concentrated in the surface horizons of most soils and are responsive to ambient vegetation. It was not possible to assemble a large data base relative to land use changes, so we used pedons from the National Soil Survey Laboratory data base. Data from classified pedons were ordered according to soil temperature regimes. Mean values for the amount of organic C per square meter to a depth of 30 cm are presented in Fig. 6–4. The *x*-axis reflects the midpoint of each soil temperature family, i.e., mesic families are defined as 8 to 15 °C mean annual soil temperature, thus, the midpoint is 11.5 °C. Isotemperature regimes reflecting tropical conditions of ≤5 °C temperature variation in the soil from summer to winter (Soil Survey Staff, 1975) are plotted separately from the temperate zone temperature regimes. Standard deviation values for each mean are plotted.

To calculate an expected change in the soil organic C pool with a global warming of 3 °C, the slope of the midpoints of the frigid and hyperthermic (non-iso) soil temperatures was used. The "iso" soil temperature regime values were not calculated because the data base is small and significant warming in the tropics is not considered to occur in the current climatic estimates. No effort was made to determine the number of cultivated vs. noncultivated samples, but they were assumed to be comparable.

The midpoint values of frigid and hyperthermic soils differ by 21 °C, and a reduction of 4.04 kg organic C m^{-2} to a depth of 30 cm is calculated. Proportioning this amount by 3/21 to represent a 3 °C MAST increase yields a loss of 0.576 kg C m^{-2} to a depth of 30 cm with a 3 °C mean annual temperature increase. This represents an 11% decrease in soil C content to a depth of 30 cm in the "average" soil of the temperature regions. By extrapolating this rate of soil organic C decline to the land area of the temperate zone (101×10^{12} m^2) and adding it (58×10^{15} g C) to the present ambient CO_2 content of the Earth's atmosphere (736×10^{15} g C based on

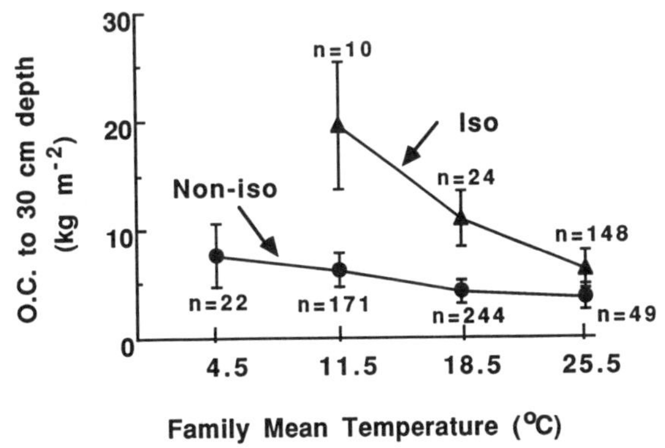

Fig. 6–4. Organic C (O.C.) content in the top 30 cm of soils by soil temperature family. Iso = soils where mean summer and mean winter soil temperatures differ by less than 5 °C; Non-iso = soils where mean summer and mean winter soil temperatures differ by 5 °C or more.

a CO_2 concentration of 350 μmol mol^{-1}),[2] we calculated that the present CO_2 content of the atmosphere would increase by approximately 8% during a 50-yr period.

Several sources of error are associated with this projection. First, the standard deviations in Fig. 6–4 are great; thus, confidence in midpoint values can be questioned. Albeit, this calculation assumed no change in organic input to the soil, and increased atmospheric CO_2 concentrations may increase plant growth and thereby increase the rate of annual organic residue return to the soil (Goundriaan & Unsworth, Chapter 8, this book).

To estimate the amount of increased annual biomass production needed to arrest the rate of soil organic C decline, we divide the projected organic C decline by 50 yr to obtain an annual rate of 115 kg C ha^{-1} yr^{-1}. Assuming a 45% efficiency of conversion of organic residues to soil organic matter and a 45% C content of dry matter, an increase of 568 kg ha^{-1} yr^{-1}, total above- and belowground dry matter, would result in no change in soil organic C contents. Clearly these data indicate more refinement is needed before more confidence can be placed in the predictions; nevertheless, a comparison of this gross value with predicted increases in biomass productivity in the temperate region with increasing CO_2 concentration appears relevant.

Carbon/Nitrogen Ratios and Soil Nitrogen Status

Higher soil temperatures are related to lower C/N ratios (Jenny, 1941). Using the same organic C values determined from mesic and thermic soils, a C/N ratio reduction of 3.1 per 3 °C temperature increase is indicated from a review of published data (Jenny, 1980). Values are highly variable and very dependent on type of vegetation. The predicted C reduction from a median 5.041 to 4.465 kg C m^{-2} for "temperature" soils coupled with a 3.1 C/N decrease from a 16/1 C/N with a 3 °C temperature rise would result in 310 kg N ha^{-1} being incorporated into the soil. This represents a 10% increase in soil N contents in the average temperate soil. If this warming process takes place over a 50-yr period, this would result in an annual increase of only 6 kg N ha^{-1} yr^{-1}.

Soil Fertility Aspects

Modern agricultural production depends more on fertilizer inputs and not on native soil fertility. Thus, any change in natural soil fertility is of minor concern in the highly developed temperature region. This is a relatively recent relationship, perhaps only recognizable in the USA since World War II. An indication of the magnitude of this shift away from dependency on native fertility in the USA can be seen from the data in Tables 6–3 and 6–4. In the early 1920s, fertilizer use for corn was a standard farm practice on the Ultisols of southeastern USA. During the same period, inorganic fer-

[2] Estimates of atmospheric CO_2 contents vary depending on the changing CO_2 concentration: see Bolin et al., 1979; Stumm & Morgan, 1981; Houghton & Woodwell, 1989.

tilizer was not used on the Mollisols in the Corn Belt of the Midwest. Corn yields on the unfertilized Mollisols were higher than those reported on the fertilized Ultisols of the southeastern USA. By the early 1980s with much higher yields in both areas, fertilizer use was higher in the midwestern Mollisols than in the southeastern Ultisols, indicating little or no dependance on native fertility.

Although the native fertility disadvantage of the Ultisols may be attributed in part to the higher rainfall and temperatures, this is not a totally valid relationship; parent material is undoubtedly a very significant soil-forming factor contributing to differences in native fertility between Mollisols and Ultisols. Albeit, any natural fertility advantage of the Mollisols seems largely lost in view of commercial farming practices.

From the examples presented in Tables 6–3 and 6–4, and from our understanding of farming practices throughout the world, we concluded that the amount of fertilizer required is directly determined by (i) amount of N, P, and K removed from the field by the harvest of the crop; and (ii) the efficiency of fertilizer use by the crop. Temperature increases of the magnitude predicted do not appear likely to affect P fixation. Potassium can leach in sandy, low effective cation exchange capacity plow layers. However, movement below the rooting zone has not been reported to our knowledge, if subsoils contain more clay than the plow layer.

Nitrogen can be lost by denitrification and translocation below rooting depth. Increased rainfall can contribute to both denitrification by raising water tables on poorly drained soils and leaching in sandy, well-drained soils. In attempting to predict the potential influence of increasing or decreasing rainfall by 10%, many confounding factors are encountered. Numerous studies have reported a high proportion of the added N fertilizer as denitrified or lost. Most of the studies, regardless of location, rainfall, or soil type, tend to cluster around a 50% recovery rate for added N. Time of fertilizer application is often critical to N fertilizer loss, and a common practice is to apply a larger proportion of the N fertilizer as a banded, side-dress application just before the most rapid vegetative growth of the crop is expected. This practice appears to be more critical in sandy soils in higher rainfall areas. Thus, we conclude that any climatic change involving increased rainfall would not result in a change in amount of N fertilizer required, but may result in some alteration in the method and timing of application.

Lime Requirement

Except for soils with carbonate as a substantial mineralogical component, all soils tend to become acid in their surface horizons. This comes about naturally by leaching and acidification from organic matter decomposition. Increased rainfall, with increased water leaching through the plow layer, tends to increase the rate of acidification, and thus the frequency and/or the amount of lime required. Our attempts to quantify additional lime requirement that would be expected to result from increased rainfall are complicated by regional differences in soil mineralogy, specific cultivar tolerance to Al in the soil

Table 6-3. Comparison of fertilizer use and yields in representative Ultisols in North Carolina.

Crop variables§	Wilson County† (Norfolk soil) 1925 Report	Wilson County† (Norfolk soil) 1983 Report	Vance County‡ (Appling soil) 1918 Report	Vance County‡ (Appling soil) 1980 Report
	kg ha^{-1}			
Corn yield	1568–2509	6899	1254–2509	5958
Tobacco yield	896–1120	3360	560–1344	2688
Cotton yield	420–560	NG	280–560	NG
Corn fertilizer rates	18–36 N (at plant) 18–27 N (sidedress) 7–13 P_2O_5 7–13 K_2O	22 N (at plant)# 112–157 N (sidedress) 45 P_2O_5 90 K_2O	34–56 (at plant) 10–27 P_2O_5 7–13 K_2O	28 N (at plant)# 112–157 N (sidedress) 112 P_2O_5 56 K_2O
Tobacco fertilizer rates	71–90 N 27–34 P_2O_5 27–34 K_2O	39–45 N (at plant)# 17–28 N (sidedress) 34–56 P_2O_5 112–134 K_2O	71–90 N 27–34 P_2O_5 18–22 K_2O	39–45 N (at plant)# 17–22 N (sidedress) 34–56 P_2O_5 112–134 K_2O

† Wilson County—Typic Paleudult; fine-loamy, siliceous, thermic: Norfolk series. 91 681 ha in the county. 1925, 36 745 ha cropped; 1978, 38 475 ha cropped (Jurney & Davis, 1925; Sink, 1983).

‡ Vance County—Typic Hapludult; clayey, kaolinitic, thermic: Appling series. 64 540 ha in the county. 1909, 12 660 ha cropped; 1971, 12 960 ha cropped (Perkins & Davis, 1921; Stimpson et al., 1980).

§ Tobacco, *Nicotiana tabacum* L.; cotton, *Gossypium hirsutum* L.

¶ NG = not grown in significant acreage.

Baird et al., 1979.

Table 6-4. Comparison of fertilizer use and yields in representative Mollisols in Iowa.†

	Muscatine soil		Tama soil	
Crop variables‡	1919	1979	1919	1979
	kg ha^{-1}			
Corn yield	2509–5645 (avg. 2634)	8091–8216	"Yields slightly lower than Muscatine"	7526–7965
Oat yield	1004–2150	3441–3412		3154–3405
Barley yield	1200–3352	NG		NG
Hay yield	2240–4480	12 400		10 900–11 600
Wheat yield	1008–1680	NG		NG
Alfalfa yield	4480–8960	NG		NG
Soybean yield	NG§	3392–3360	NG	2957–3226
	1919	1982#, ††	1919	1982#
Corn fertilizer rates	"Commercial fertilizer never used"¶	168–202 N 78–123 P_2O_5 90–134 K_2O	"Commercial fertilizer never used"	168–202 N 78–123 P_2O_5 90–134 K_2O
Oat		82 N 45 P_2O_5		76 N 45 P_2O_5 78 K_2O

† Cedar County—Aquic Argiudoll; fine-silty, mixed, mesic: Muscatine series. Typic Argiudoll; fine-silty, mixed, mesic: Tama series. 147 744 ha (1910) with 141 982 ha in farms; 151 632 ha (1974) with 142 911 ha in farms (O'Neal & Gray, 1919; Schumerhorn et al., 1979).

‡ Oat, *Avena sativa* L.; barley, *Hordeum vulgare* L.; alfalfa, *Medicago sativa* L.; soybean, *Glycine max* (L.) Merr.

§ NG = not grown in significant acreage.

¶ Alfalfa was limed and inoculated.

\# Voss, 1982.

†† Medium soil test level Ag-65 (Rev.) Iowa State Univ., Ames, IA.

solution, and contrasting methods of lime requirement determination. Our opinion is that a rainfall increase of 10% would result in an almost undetectable increase in lime requirement in most agricultural areas. Also, the development of Al-tolerant food crop cultivars indicates that lime requirement in many areas may be reduced.

CONCLUSIONS

An increase in mean annual air temperatures will result in a corresponding increase in mean annual soil temperature. Although warming is expected to increase solubility, generally increasing the solubility of soil minerals, an increase of 3 °C is too small to engender measurable change in most soil chemical or mineralogical properties. Increased soil temperature can be expected to decrease the amount of soil organic matter and result in CO_2 release to the atmosphere unless annual dry matter production increases. Soil C/N ratios are expected to narrow. The soil temperature change can be expected to move traditional corn and wheat farming further north in the Northern

Hemisphere but have little or no effect on overall rates of fertilizer use. Timing and method of fertilizer application may be slightly modified if both temperature and rainfall patterns are altered.

REFERENCES

Baird, J.V., F.R. Cox, E.J. Kamprath, S.N. Hawks, E.L. Kimbrough, and G.S. Miner. 1979. Fertilizer suggestions for field, pasture, and hay crops. p. 343–356. *In* The 1979 North Carolina agricultural chemicals manual. N.C. State Univ., Raleigh, NC.

Bigham, J.M., D.C. Golden, S.W. Buol, S.B. Weed, and L.H. Bowen. 1978. Iron oxide mineralogy of well-drained Ultisols and Oxisols: II. Influence on color, surface area, and phosphate retention. Soil Sci. Soc. Am. J. 42:825–830.

Bolin, B., E.T. Degens, P. Duvigneaud, and S. Kempe. 1979. The global biogechemical carbon cycle. p. 1–56. *In* B. Bollin et al. (ed.) The global carbon cycle, scope 13. John Wiley & Sons, New York.

Food and Agricultural Organization of the United Nations. 1975. Soil map of the world. Vol. 2: North American. UNESCO, Paris, France.

Food and Agriculture Organization of the United Nations. 1978a. Soil map of the world. Vol. 5: Europe. UNESCO, Paris, France.

Food and Agriculture Organization of the United Nations. 1978b. Soil map of the world. Vol. 8: North and Central Asia. UNESCO, Paris, France.

Food and Agriculture Organization of the United Nations. 1985. Production yearbook. FAO, Rome, Italy.

Garrels, R.M., and C.L. Christ. 1965. Solutions, minerals, and equilibria. Harper & Row, Publishers, New York.

Hargreaves, G.H., and Z.A. Samani. 1986. World water for agriculture—precipitation management. Int. Irrig. Ctr. Dep. of Agric. and Irrig. Eng. Utah State Univ., Logan, UT.

Houghton, R.A., and G.M. Woodwell. 1989. Global climate change. Sci. Am. 260:36–44.

Jenny, J. 1941. Factors of soil formation. McGraw-Hill, New York.

Jenny, H. 1980. The soil resource. Springer-Verlag New York, New York.

Jurney, R.C., and W.A. Davis. 1925. Soil survey of Wilson County, N.C. USDA Bur. of Chem. and Soils, U.S. Gov. Print. Office, Washington, DC.

Kampf, N., and U. Schwertmann. 1982. Goethite and hematite in a climosequence in southern Brazil and their application in classification of kaolinitic soils. Geoderma 29:27–39.

O'Neal, A.M., Jr., and D.A. Gray. 1919. Soil survey of Cedar County, Iowa. p. 1427–1457. *In* USDA Bur. of Soils, 21st Rep., U.S. Gov. Print. Office, Washington, DC.

Perkins, S.O., and W.A. Davis. 1921. Soil survey of Vance County, N.C. USDA Bur. of Soils, U.S. Gov. Print. Office, Washington, DC.

Schumerhorn, E.J., L.I Harmon, F.F. Riecken, and T.E. Fenton. 1979. Soil survey of Cedar County, Iowa. USDA-SCS, U.S. Gov. Print. Office, Washington, DC.

Schwertmann, U., E. Murad, and D.G. Schulze. 1982. Is there Holocene reddening (hematite formation) in soils of a xeric temperature areas? Geoderma 27:209–223.

Sink, L.T. 1983. Soil survey of Wilson County, N.C. USDA-SCS, U.S. Gov. Printing Office, Washington, DC.

Smith, G.D., F Newhall, and L.H. Robinson, and D. Swanson. 1964. Soil-temperature regimes: Their characteristics and predictability. SCS Tech. Publ. 144. USDA-SCS, Washington, DC.

Soil Survey Staff. 1975. Soil taxonomy. USDA Handb. 436. U.S. Gov. Print. Office, Washington, DC.

Sposito, G., and S.V. Mattigod. 1979. Geochem: A computer program for the calculation of chemical equilibria in soil solutions and other natural solutions. Dep. of Soil and Environ. Sci., Univ. of California, Riverside, CA.

Stimpson, J.V., R.L. Mathis, Jr., W.I. Shope, J.W. Cawthorn, and D. Clapp. 1980. Soil survey of Vance County, N.C. USDA-SCS, U.S. Gov. Print. Office, Washington, DC.

Stumm, W., and J.J. Morgan. 1981. Aquatic chemistry. John Wiley & Sons, New York.

Voss, R.D. 1982. General guide for fertilizer recommendations in Iowa. AG-65 (Rev.) Iowa State Univ., Ames, IA.

7 Implications of Increasing Carbon Dioxide and Climate Change for Plant Communities and Competition in Natural and Managed Ecosystems

David T. Patterson
USDA-ARS
Duke University
Durham, North Carolina

Elizabeth P. Flint
Duke University
Durham, North Carolina

ABSTRACT

Future increases in the CO_2 concentration of the Earth's atmosphere will directly affect the physiological processes and growth of plants. Indirect climatic effects, including global warming and changes in precipitation patterns and the frequency of weather extremes, may have greater impact than direct effects of CO_2 on physiological processes. Carbon dioxide enrichment up to twice ambient levels or more generally increases plant growth, although the magnitude of growth stimulation varies greatly with species, photosynthetic pathway, growth stage, and water and nutrient status. In both natural and managed ecosystems, differential growth responses to both CO_2 concentration and climatic change will affect future competitive ability and fitness of plants. The relative importance of various weed species in agroecosystems may change, but selection of adapted crop varieties and management methods may minimize negative impacts. In natural ecosystems, species extinctions probably will increase, because migration and adaptation through natural selection may be too slow to accommodate the rapid climatic changes involved. Weedy species with broad ecological amplitudes are likely to prosper at the expense of endemic species or those already in marginal habitats.

 Impact of Carbon Dioxide, Trace Gases, and Climate Change on Global Agriculture, ASA Special Publication no. 53.

Current and projected increases in the global atmospheric CO_2 concentration have led to widespread concerns about long-term effects on the Earth's climate and vegetation. The atmospheric CO_2 concentration already has increased from a preindustrial level of approximately 280 μmol mol^{-1} to the current 345 to 350 μmol mol^{-1} value (Schlesinger & Mitchell, 1985). Because of accelerating use of fossil fuels and conversion of forested land to agricultural production, the atmospheric CO_2 concentration is expected to continue to increase to 600 μmol mol^{-1} by the middle to end of the 21st century (Trabalka et al., 1985).

Atmospheric CO_2 is a major resource for plant growth. Carbon from CO_2 fixed through photosynthesis constitutes about 45 to 50% of plant dry matter (Houghton et al., 1985). Increased concentrations of CO_2 in the Earth's atmosphere will increase photosynthesis in most C_3 plants, for which current CO_2 levels are suboptimum. There are many other direct physiological effects of increased CO_2, which will be considered later in the context of plant competition and communities.

In addition to affecting plants directly, increasing atmospheric CO_2 also may affect the Earth's climate (Taylor & MacCracken, Chapter 1, this book). Because CO_2 is a selective filter for radiation, it allows shortwave solar radiation to reach the Earth's surface, but absorbs and reradiates the longwave thermal radiation from the Earth's surface (Bolin et al., 1986). In addition to CO_2, other "greenhouse" gases, such as CH_4, N_2O, and chlorofluorocarbons, are radiatively active and may also contribute to global warming (Burke & Lashof, Chapter 3, this book). The concentrations of these trace gases appear to be increasing more rapidly than that of CO_2. The combined effects of CO_2 and these other greenhouse gases could accelerate global warming to the equivalent of a doubling of the CO_2 concentration by 2040 (Bolle et al., 1986).

There is considerable controversy concerning the magnitude of the so-called greenhouse effect on the Earth's equilibrium surface temperature (Luther & Cess, 1985; Idso, Chapter 2, this book), but the consensus seems to be that the Earth's temperature will increase as the concentration of radiatively active gases rises. Three recent general circulation models (GCM) project increases in the range of 3.5 to 4.2 °C for mean global temperature as a response to an equivalent doubling of the global CO_2 concentration (Schlesinger & Mitchell, 1985). The increases would be greater in winter than in summer and greater at high latitudes than at the equator (Dickinson, 1986).

On a global scale, the magnitude of the expected climate change by the mid-21st century will be comparable to that between the peak of the last glaciation, 18 000 yr ago, and the present (Pittock, 1988). A global temperature change of this magnitude probably would be accompanied by significant-changes in precipitation (both amount and frequency), wind patterns, and potential evapotranspiration. Regional projections of these variables are only speculative, based on present knowledge (Dickinson, 1986). Manabe and Wetherald (1986) projected significant reductions in summer soil moisture in large areas of the middle to high latitudes of the Northern Hemisphere. For example, in the U.S. Great Plains, summer soil moisture decreased 30 to 50% in a GCM simulation based on a doubling of atmospheric CO_2 and a consequent increase in mean surface air temperature of about 4 °C. In a more complex model simulation that incorporated seasonal ocean heat storage, Washington and Meehl (1984) reported decreased soil moisture in tropical and subtropical continental areas and increased soil moisture at high latitudes. In contrast to Manabe and Wetherald (1986), they also projected increased soil moisture early in the growing season in midlatitudes. On a more localized scale, Coleman (1988) has used past weather data to project decreased summer precipitation in peninsular Florida in response to global warming.

Thus far climatic change has been discussed in terms of alteration of regional means for temperature and precipitation. Mean values set general limits for ecosystem productivity and determine large-scale species distribution (Graetz, 1988). But the effects of climatic change must also be evaluated in terms of the variability associated with the climatic means, and in terms of the frequency of extreme climatic events. Such extreme events exert a disproportionate effect on the population processes that determine structure and species composition of plant communities, especially those dominated by perennials (Graetz, 1988).

The predicted CO_2-induced changes in mean temperature and precipitation are likely to be accompanied by considerable increases in the frequencies of extreme climatic events. Parry and Carter (1985) have demonstrated that the frequency of such climatic events is highly sensitive to relatively small changes in the mean. For example, if mean annual precipitation in a location decreases by just one standard deviation, then droughts of a magnitude that formerly occurred only once per century would be expected to recur about every 11 yr (Wigley, 1985). Along with increased probabilities of floods, droughts, and temperature extremes, CO_2-generated global warming may also increase the probability of severe winds and strong convective storms, including tropical cyclones (Pittock, 1988). The probabilities of fire may also increase in regions where increased CO_2 reduces the mean relative humidity (Beer et al., 1988).

Furthermore, in any situation where changes in the mean increase the frequencies of extreme events, the probability of two successive extreme events automatically increases by an even greater factor (Wigley, 1985). Thus, reduced mean precipitation will be associated with greatly increased frequency of consecutive severe droughts. Consecutive extreme events may occur on the scale of years (e.g., droughts), or on a scale of days within the growing season. Simulations by Mearns et al. (1984) indicate that a small rise in mean growing season temperatures in the U.S. Corn Belt greatly increases the probability that corn (*Zea mays* L.) plants will be subjected to several consecutive days of temperatures high enough to significantly reduce yield. As Parry and Carter (1985) have pointed out, farmers and agricultural systems, especially those operating in ecologically or economically marginal situations, are especially vulnerable to consecutive low-yield years. Furthermore, in unmanaged ecosystems, the very survival of many plants that can tolerate a single extreme event may be threatened by the occurrence of consecutive extreme events.

DIRECT EFFECTS OF INCREASED CARBON DIOXIDE CONCENTRATION

There is more agreement on the direct physiological effects of increasing CO_2 on plants than on the effects of CO_2 on global climate. Physiological effects are discussed elsewhere (Acock, Chapter 4; Idso, Chapter 5; Goudriaan & Unsworth, Chapter 8, this book) and will be reviewed only briefly here. Primary physiological effects include increased net photosynthesis in C_3 plants, reduced photorespiration, changes in dark respiration, and reduced stomatal conductance (Gifford, 1988; Gifford et al., 1985). Secondary effects include decreased transpiration; increased tolerance of air pollution; increased water use efficiency; alterations in composition, concentration, and translocation of photosynthate; and increases in the C/N ratio of plant tissue (Strain, 1987). Integrated effects at the whole-plant level include increased dry matter production, increased leaf area, changes in leaf structure, modifications in growth form, and alterations in developmental phenology, reproduction, and senescence (Kramer,1981; Strain,1987).

Most of these reported effects have implications for levels of organization above that of the individual plant. For example, changes in growth form and growth potential may influence competitive ability, as may increases in water use efficiency. Increased photosynthate supply may enhance symbiotic N_2 fixation (Lamborg et al., 1983). Increased C/N ratios of plant tissue reduce nutritional quality (Overdieck et al., 1988) and therefore increase feeding by herbivorous insects (Lincoln et al., 1984, 1986). Increased C/N ratios in plant litter may slow decomposition by microorganisms and therefore interfere with nutrient cycling. Enhanced production of so-called "secondary metabolites" may increase the potential for allelopathic interactions among plants (Strain, 1985).

Most of the primary and secondary physiological effects of the increases in CO_2 concentration projected for the next several centuries probably will be beneficial to plants grown in monoculture in managed ecosystems. There is increasing consensus that CO_2 enrichment can increase plant growth even when light, water availability, or most nutrients are limiting (Kimball, 1986a, b; Patterson & Flint, 1982; Sionit et al., 1981a, b; Warrick et al., 1986b; Goudriaan & Unsworth, Chapter 8, this book). However, P deficiency may eliminate growth enhancement (Barlow & Conroy, 1988; Goudriaan & de Ruiter, 1983). Low temperatures appear to reduce or eliminate high-CO_2 growth enhancement (Hofstra & Hesketh, 1975; Sage & Sharkey, 1987; Idso et al., 1987; Idso, Chapter 5, this book). On the other hand, CO_2 enrichment may increase plant tolerance to temperature extremes (Baker et al., 1989; Sionit et al., 1981b; Sionit et al., 1987; Sasek & Strain, 1988).

Recent reviews by Kimball (1983) and Cure (1985) indicate that, with few exceptions, CO_2 enrichment to at least two or three times current ambient CO_2 levels stimulates the growth of plants in pots in controlled environments. Kimball (1983) reported biomass increases in response to doubled ambient CO_2, which ranged from 1.10 to 2.43$\times$ control values in C_3 crop plants, and from 0.98 to 1.24$\times$ in C_4 crop plants. At the time of that review

(1983), few studies of the effects of CO_2 enrichment on noncrop plants were available. The few data then available for woody plants and noncrop herbaceous plants indicated response ranges similar to those of crop plants.

The data from a number of more recent studies of noncrop species have been extracted and assembled in Table 7–1. The responses of most of these noncrop plants fall into the same ranges originally reported by Kimball (1983), but there is a notable difference. In agricultural weeds and native herbaceous species, the range of response is quite similar for plants with the C_3 and C_4 photosynthethic pathways. The response ranges of the C_3 crops were quite different from those of the C_4 crops in Kimball's survey, although it included relatively few studies of C_4 plants. Furthermore, researchers have reported large differences in growth response to CO_2 enhancement within some species, particularly those with C_4 photosynthesis. Some inconsistencies in the reported responses of noncrop species to CO_2 enrichment may be related to differences in growth conditions. For example, Patterson (1986) reported that C_4 grasses responded dramatically to CO_2 enrichment when growth was limited by water availability. However, genetic variability among individuals and populations probably also contributes to differential within-species responses to CO_2 enrichment. The studies of *Phlox drummondii* Hook. (C_3) by Garbutt and Bazzaz (1984) and of barnyardgrass [*Echinochloa crusgalli* (L.) Beauv.] (C_4) by Potvin and Strain (1985) demonstrate that different accessions of noncrop species may respond differently to increased CO_2 concentration even when all other environmental variables are controlled.

The surveys of Kimball (1983) and Cure (1985) demonstrate that even among plants sharing the same photosynthetic pathway, considerable potential exists for different growth responses to increased atmospheric CO_2 concentrations. These differences in growth response will influence plant competition and other processes of importance in both managed and natural ecosystems.

EFFECTS OF CARBON DIOXIDE CONCENTRATION ON PLANT COMPETITION

Competition is the mutual struggle for the limited resources upon which plant growth and reproductive success depend. Competition occurs when one individual reduces the availability of a resource below a level at which it limits the growth or reproductive performance of a neighbor. Any factor, such as CO_2 concentration, that affects the growth of an individual in turn affects its ability to compete with neighbors of the same or different species. According to Grime (1979), competition is probably most severe in habitats with relatively low disturbance and a moderate physical and chemical environment. In highly disturbed or environmentally harsh sites, plant density may be too low for significant competition to occur.

In many natural ecosystems, competition is a major determinant of vegetation dynamics and species composition. Intraspecific competition influences

Table 7-1. Effects of doubling CO_2 concentration on biomass of noncrop plants.

Plants	Range of response (× growth at ambient)	Number of measurements	References†
	Herbaceous C_3 species		
Abutilon theophrasti Medik.	1.23–1.44	5	1, 5, 11, 14, 32
Agoseris heterophylla (Nutt.) Greene	1.60	1	29
Agropyron smithii Rydb.	1.31	1	24
Ambrosia artemisiifolia L.	1.10–2.00	2	1, 32
Anoda cristata (L.) Schlecht	1.40	1	14
Aster pilosus Willd.	1.15–2.10	2	30, 31
Bromus mollis L.	1.37	1	6
Bromus tectorum L.	1.54	1	24
Carex bigelowii Torr.	1.13	1	9
Cassia obtusifolia L.	1.60	1	12
Chenopodium album L.	1.22–1.55	2	2, 32
Clarkia rubicunda Lewis & Lewis	1.80	1	26
Crotalaria spectabilis Roth	1.67	1	12
Datura stramonium L.	1.74–2.72	3	1, 2, 5
Eichornia crassipes (Mart.) Solms	1.32	1	25
Festuca elatior L.	1.32	1	3
Lasthenia glabrata Lindl.	0.79	1	29
Layia platyglossa (F. & M.) Gray	1.43–1.54	2	26, 29
Lolium perenne L.	1.43	1	7
Micropus californicus (F. & M.)	1.11	1	29
Microseris sp.	0.90	1	29
Nemophila menziesii H. & A.	1.03	1	26
Oryzopsis hymenoides Ricker	1.10	1	24
Phlox drummondii Hook.	1.04–2.06	4	5
Plantago erecta Morris	1.03	1	29
Plantago major L.	1.55	1	15
Polygonum pensylvanicum L.	1.11–1.53	2	2, 32
	Herbaceous C_4 species		
Amaranthus retroflexus L.	0.90–1.41	3	1, 2, 32
Andropogon virginicus L.	0.84–1.17	2	30, 31
Bouteloua gracilis Lag. ex Steud.	1.36	1	18
Digitaria sanguinalis (L.) Scop.	1.06–1.61	2	10, 23
Echinochloa crusgalli (L.) Beauv.	0.95–1.59	3	10, 16, 23
Eleusine indica (L.) Gaertn.	1.02–1.21	3	10, 16, 23
Eragrostis orcuttiana Vasey	2.50	1	24
Rottboellia exaltata L.f.	1.21	1	11
Setaria faberii Herrm.	0.90–1.42	4	2, 23, 32
Setaria lutescens Hubb.	1.70–1.74	2	2
Sorghum halepense (L.) Pers.	0.56–1.10	2	3, 13
	Woody species		
Acer saccharinum L.	1.04–1.61	2	1, 28
Betula nana L.	0.90	1	9
Carya ovata (Miller) K. Koch	1.00	1	28
Cecropia obtusifolia	1.14	1	17
Citrus sinensis (L.) Osbeck‡	1.26	1	4
Fraxinus lanceolata Borkh.	0.87	1	28
Ledum palustre L.	0.90	1	9
Liquidambar styraciflua L.	1.33–1.73	3	19, 22, 27
Liriodendron tulipifera L.	0.94	1	28

(continued on next page)

Table 7-1. Continued.

Plants	Range of response (× growth at ambient)	Number of measure-ments	References†
Lonicera japonica Thunb.	2.10	1	21
Myriocarpa longipes	0.92	1	17
Ochroma lagopus Swartz	1.79	1	8
Pentaclethra macroloba (Willd.) Kuntze	1.30	1	8
Pinus taeda L.	0.93–1.59	3	19, 22, 27
Piper auritum	1.15	1	17
Platanus occidentalis L.	1.07–1.13	2	1, 28
Populus deltoides Bart.	1.65	1	1
Pueraria lobata (Willd.) Ohwi	1.20	1	20
Quercus rubra L.	1.28	1	28
Senna multijuga	1.20	1	17
Trichospermum mexicanum	1.20	1	17

† Key to references: (1) Carlson & Bazzaz, 1980, (2) Carlson & Bazzaz, 1982, (3) Carter & Peterson, 1983, (4) Downton et al., 1987, (5) Garbutt & Bazzaz, 1984, (6) Larigauderie et al., 1988, (7) Nijs et al., 1988, (8) Oberbauer et al., 1985, (9) Oberbauer et al., 1986, (10) Patterson, 1986, (11) Patterson & Flint, 1980, (12) Patterson & Flint, 1982, (13) Patterson et al., 1984, (14) Patterson et al., 1988, (15) Poorter et al., 1988, (16) Potvin & Strain, 1985, (17) Reekie & Bazzaz, 1989, (18) Riechers & Strain, 1988, (19) Rogers et al., 1983, (20) Sasek & Strain, 1988, (21) Sasek & Strain, 1989, (22) Sionit et al., 1985, (23) Sionit & Patterson, 1984, (24) Smith et al., 1987, (25) Spencer & Bowes, 1986, (26) St. Omer & Horvath, 1983, (27) Tolley & Strain, 1984, (28) Williams et al., 1986, (29) Williams et al., 1988, (30) Wray & Strain, 1986, (31), Wray & Strain, 1987a, (32) Zangerl & Bazzaz,1984.

‡ High CO_2 was 800 μmol mol^{-1}.

the growth, fecundity, and mortality of individuals. Interspecific competition affects not only these characteristics of individual plants, but also influences the distribution of species on a local scale. For example, interspecific competition may restrict a species to microsites or soil types that will not support the growth of its more aggressive competitors (Begon et al.,1986).

Competition also is important in managed ecosystems. In most crop monocultures, intraspecific competition influences optimum planting density and final yield per unit area. Interspecific competition between crops and the weeds that usurp resources needed by the crops is a major cause of crop yield loss.

At a meeting in 1982 to discuss direct effects of the CO_2 increase on terrestrial plant communities, no published studies of the effects of CO_2 enrichment on plant competition could be identified, though several experiments were in progress (Strain & Bazzaz, 1983). Subsequently, a number of studies have been conducted and published (Table 7–2). Most of these deal with interspecific competition, but a recent design by Bazzaz and Garbutt (1988) incorporates intraspecific competition as well. Most of the studies have utilized annual or short-lived perennial species that are important both in managed ecosystems as agricultural weeds and in natural ecosystems as pioneers of secondary succession.

Because of demonstrated differences in the response of C_3 and C_4 plants to CO_2 enrichment, several researchers have compared the perfor-

Table 7-2. Summary of experiments testing competition among plants exposed to different CO_2 levels.†

Number of species competing	Species involved	P_s, pathway of species	CO_2 conc. levels, μmol mol^{-1}	Other variables tested	Species favored by increased CO_2	Effect of increased CO_2 on community production	Other responses of increased CO_2 conc.; interactions	Source
2	*Festuca elatior* *Sorghum halepense*	C_3 C_4	350, 600	None	C_3	0	Increased competitiveness of C_3 species; community productivity unchanged.	Carter & Peterson, 1983
2	*Glycine max* *Sorghum halepense*	C_3 C_4	350, 675	None	C_3	+	Increased competitiveness or C_3 species; RYT unchanged.	Patterson et al., 1984
2	*Aster pilosus* *Andropogon virginicus*	C_3 C_4	350, 500, 650	Planting date (separate exp.)	C_3	0	Increased competitiveness of C_3 species if planted together; delay in planting the C_3 species offset its competitive advantage in increased CO_2.	Wray & Strain, 1987a (same planting date) Wray & Strain, 1987b (delayed planting date)
2	*Aster pilosus* *Andropogon virginicus*	C_3 C_4	350 650	Water status	C_3	+	Increased competitiveness of C_3 species regardless of water status.	Marks & Strain, 1989
2	*Abutilon theophrasti* *Amaranthus retroflexus*	C_3 C_4	350, 500, 700	None	C_4 (growth); C_3 (P_s)	+	Increased competitiveness of C_4 species; greater P_s enhancement in C_3 but greater RGR increase in C_4 due to changes in C allocation.	Bazzaz et al., 1989
2	*Lolium perenne* *Trifolium repens*	C_3 C_3	340, 547 yr.1 340, 600 yr.2	None (compare with exp. below)	*Trifolium*	+	Community production up (greater effect in first year than later; both spp. benefited early in exp.; *Trifolium* favored over long term.	Overdieck et al., 1984

(continued on next page)

Table 7–2. Continued.

Number of species competing	Species involved	P_s, pathway of species	CO_2 conc. levels, $\mu mol\ mol^{-1}$	Other variables tested	Species favored by increased CO_2	Effect of increased CO_2 on community production	Other responses of increased CO_2 conc.; interactions	Source
2	*Lolium perenne* *Trifolium repens*	C_3 C_3	300, 620	Four mowings yr^{-1} to simulate pasture management	Variable	+	Community production up; *Trifolium* favored before mowing, but *Lolium* more competitive after mowing.	Overdieck & Reining, 1986
4	*Abutilon theophrasti* *Ambrosia artemisiifolia* *Polygonum pensylvanicum* *Amaranthus retroflexus*	C_3 C_3 C_3 C_4	300, 600, 1200	Soil moisture; (dry/wet)	C_3 spp.	+ (at both levels of soil moisture)	C_3 species favored, but responses nonlinear and variable among species; CO_2 effects modified by soil moisture; community biomass up at both soil moistures.	Bazzaz & Carlson, 1984
4	*Abutilon theophrasti* *Ambrosia artemisiifolia* *Amaranthus retroflexus* *Setaria faberii*	C_3 C_3 C_4 C_4	350, 500, 700	None	*Abutilon* most favored; *Setaria* least favored; others intermediate	0	CO_2 response was specific to species; responses to increased CO_2 nonlinear; no increase in community biomass or community seed production. Significant interaction effect of number of competing species.	Bazzaz & Garbutt, 1988
6	*Abutilon theophrasti* *Ambrosia artemisiifolia* *Chenopodium album* *Polygonum pensylvanicum* *Amaranthus retroflexus* *Setaria faberii*	C_3 C_3 C_3 C_3 C_4 C_4	300, 450, 600	Light: (30%/100% of full); Nutrient: (low, med., high)	C_3 spp.	Increased as CO_2 increased 300 to 450 $\mu mol\ mol^{-1}$ but no change from 450 to 600	C_3 species favored, but responses nonlinear and variable among species; CO_2 effects modified by light, nutrients; light modified community production response, but nutrients did not.	Zangerl & Bazzaz, 1984

(continued on next page)

Table 7-2. Continued.

Number of species competing	Species involved	P_s, pathway of species	CO_2 conc. levels, μmol mol^{-1}	Other variables tested	Species favored by increased CO_2	Effect of increased CO_2 on community production	Other responses of increased CO_2 conc.; interactions	Source
4	Serpentine grassland herbaceous species: *Agoseris heterophylla* *Lasthenia glabrata* *Layia platyglossa* *Micropus californicus*	All C_3	350, 500, 700	None	Variable	0	Individual species responded differentially and nonlinearly to CO_2 enrichment. Other resources apparently limited the total community production.	Williams et al., 1988
5	Tropical trees: *Cecropia obtusifolia* *Myriocarpa longipes* *Piper auritum* *Senna multijuga* *Trichospermum mexicanum*	All C_3	350, 525, 700	None	Favored: *Cecropia* *Trichospermum,* *Piper;* Neutral: *Myriocarpa* Disadvantaged: *Senna*	0	No effects on P_s or growth of individual plants, but CO_2 affected canopy architecture, which altered light availability. Authors concluded that individual species responses reflected secondary light effects.	Reekie & Bazzaz,1989
3 and 3	Temperate trees, upland: *Carya ovata* *Liriodendron tulipifera* *Quercus rubra* Temperate trees, bottomland: *Acer saccharinum* *Fraxinus lanceolata* *Platanus occidentalis*	All C_3	350, 500, 700	Light	Variable	0	Competitive interactions of species in each group were altered; upland set showed CO_2 × species interaction; bottomland set had no interaction; total community biomass not affected by CO_2 in either group of species.	Williams et al., 1986

† 0 = no response; + = increase; P_s = photosynthesis; RYT = relative yield total; RGR = relative growth rate.

mance of competing species with the different photosynthetic pathways. Examples of experiments pairing one C_3 and one C_4 species include comparisons of: C_3 crop with C_4 weed (Patterson et al.,1984), C_3 old field successional with C_4 old field successional species (Wray & Strain, 1987a, b), and C_3 vs. C_4 ruderal grasses (Carter & Peterson, 1983). In all of these studies, conducted in pots in controlled environments, the C_3 species produced proportionately more biomass than the C_4 species when grown individually under enhanced CO_2. In competition, the differences in performance of plants with the two photosynthetic pathways at high CO_2 were even greater. For example, Patterson et al. (1984) demonstrated that C_3 soybean [*Glycine max* (L.) Merr.] competed more effectively against C_4 johnsongrass [*Sorghum halepense* (L.) Pers.] at 675 μmol mol^{-1} than at 350 μmol mol^{-1} CO_2. Carter and Peterson (1983) found that raising the CO_2 concentration significantly increased the competitive ability of C_3 *Festuca elatior* L. grown with C_4 johnsongrass. The competitive ability of C_3 *Aster pilosus* Willd. vis-a-vis the C_4 broomsedge (*Andropogon virginicus* L.) also significantly increased with increasing CO_2 concentration (Wray & Strain, 1987a). In a second experiment, the same authors found that when aster (C_3) was planted 6 wk later than broomsedge (C_4), the competitive advantage of broomsedge was so great that even under CO_2 enrichment to 675 μmol mol^{-1} the aster was unable to compete effectively (Wray & Strain, 1987b).

In the studies that permitted analysis using the deWit replacement series competition methodology (Patterson et al., 1984; Wray & Strain, 1987a), no differences in relative yield total of C_3 and C_4 species were attributable to CO_2 enhancement. This indicates that increased CO_2 did not allow any of these species to exploit resources unavailable to the other. Surprisingly, in an experiment of Bazzaz et al. (1989), increased CO_2 concentration actually increased the competitiveness of a C_4 weed (redroot pigweed, *Amaranthus retroflexus* L.) grown in competition with a C_3 weed (velvetleaf, *Abutilon theophrasti* Medik.). In this case, it appeared that increased CO_2 improved the relative growth rate of pigweed, perhaps as a result of changes in allocation patterns. Even though photosynthesis in pigweed was enhanced much less by high CO_2 than photosynthesis in velvetleaf, the stimulation of its relative growth rate ultimately favored pigweed in competition. Thus, it can be seen that effects of CO_2 enrichment other than the direct enhancement of photosynthesis may favor the C_4 rather than the C_3 species. Overall, especially when multiple-species competition experiments are considered, the evidence indicates that competitive ability of C_3 species is generally improved with respect to C_4 species under high CO_2.

The above-mentioned experiments were basically short-term studies. In another two-species system, competition between two C_3 pasture species planted in concrete basin "microcosms" was followed for 2 yr. The basins were 0.73 by 0.74 m and 0.4 m deep and filled with mixed garden soil. Overdieck et al. (1984) and Overdieck and Reining (1986) tested perennial ryegrass (*Lolium perenne* L.) in competition with white clover (*Trifolium repens* L.). The species were grown individually and in 1:1 mixture ratio in uncovered basins, and in mixture only in greenhouse-sheltered basins where CO_2 concentration could be maintained at 300 or 620 μmol mol^{-1}. In the first experiment, Overdieck et al.(1984) found that both species initially showed a positive response to CO_2 enrichment; however, over the long term at high CO_2, the clover was favored to such an extent that it crowded out the ryegrass. When the same two species were grown in competition at similar levels of CO_2, but with four mowings per year (simulating pasture cutting or grazing), ryegrass outcompeted clover even at the high CO_2 concentration (Overdieck & Reining, 1986). Because of morphological differences, clover suffers more from clipping than ryegrass. As in other experiments, it is clear that CO_2 affects competitiveness, but the nature of the effect varies with species and environmental conditions. Overdieck et al. (1984) also reported that community production was enhanced greatly at first by CO_2 enrichment, but in the second year the difference in community production between ambient and high CO_2 treatments was considerably smaller.

Carbon dioxide competition experiments have not been limited to two-species systems. Bazzaz and colleagues have published a number of studies examining the effects of several CO_2 concentrations on growth of several species in monocultures and in competing assemblages. In some cases, not only CO_2 but also other environmental factors were varied. Bazzaz and Carlson (1984) compared *Polygonum pensylvanicum* L., *Ambrosia artemisiifolia* L., and *Abutilon theophrasti* Medik. (all C_3 plants) with *Amaranthus retroflexus* L. (C_4). Zangerl and Bazzaz (1984) grew the same four species, plus *Chenopodium album* L. (C_3) and *Setaria faberii* Herrm. (C_4). In both of these experiments, competitive ability of the C_3 species improved in assemblages grown in enriched-CO_2 (as compared with ambient level) atmospheres. Also, total community production increased with CO_2 concentration, but this response was strongly nonlinear. Within groups of species sharing the same photosynthetic pathway, the CO_2 growth responses were species specific, and light, water, and nutrient availability modified the CO_2 response of each species in different ways.

Williams et al. (1988) grew six serpentine grassland herbs (all C_3 species) individually and in competitive assemblages. They found that increased CO_2 concentration did not increase total community biomass, and concluded that CO_2 concentration might enhance growth minimally or not at all in systems where other resources (e.g., nutrients in serpentine environments) are strongly limiting.

The competitive neighborhood experiment of Bazzaz and Garbutt (1988) examined performance of mono-, bi-, tri-, and tetracultures of constant density for a set of four species. The species used (C_3, *A. theophrasti* Medik. and *P. pensylvanicum* L.; and C_4, *A. retroflexus* L. and *S. faberii* Herrm.) had all been included in previous studies, but in this experiment neither total community biomass nor community seed production increased as CO_2 was increased from 350 to 500 to 700 μmol mol^{-1}. However, as in the earlier experiment, C_3 species gained in competitiveness with respect to C_4 species at the higher CO_2 concentrations. This experimental design revealed significant effects on the growth of "target" plants by the number of competing species in the mixture; growth was suppressed more severely by multispecies assemblages than in bicultures. Overall, it is clear from a survey of the current literature that CO_2 enrichment is likely to increase the relative competitive ability of C_3 plants with respect to C_4 plants.

Two studies have examined the effects of CO_2 concentration on competition in multispecies assemblages of woody species (all of which have the C_3 pathway). Williams et al. (1986) compared competition among two assemblages of temperate deciduous forest trees: an upland group [*Carya ovata* (Miller) K. Koch, *Liriodendron tulipifera* L., and *Quercus rubra* L.] and a bottomland group (*Platanus occidentalis* L., *Fraxinus lanceolata* Borkh., and *Acer saccharinum* L.). Reekie and Bazzaz (1989) investigated competition among five tropical tree species. Each experiment used three CO_2 concentrations, 350 and 700 μmol mol^{-1} and an intermediate concentration (500 to 525 μmol mol^{-1}). In these two studies, CO_2 enrichment did not lead to greater community production. However, altering CO_2 levels clearly changed the competitive interactions within each set of three temperate trees, and within the set of five tropical trees. In the tropical species, the effects of CO_2 on competitive ability operated through alterations in canopy architecture in a light-limited system rather than through differential enhancement of photosynthesis. The authors concluded that increased CO_2 concentration would be likely to lead to changes in community composition, with some species increasing and others decreasing in importance.

EFFECTS OF INCREASING CARBON DIOXIDE AND CLIMATIC CHANGE ON NATURAL ECOSYSTEMS

Only in a few experiments have simple natural ecosystems been exposed to enriched CO_2. These have been primarily restricted to systems dominated by plants small enough to permit experimental manipulations of realistic microcosms of the ecosystems, both in the field and in the laboratory. For example, Drake et al. (1988) used open-top field chambers to study the effects of CO_2 enrichment on three plant communities in the Chesapeake Bay wetland ecosystem. In monospecific communities of the C_3 sedge *Scirpus olneyi* Gray, doubling the ambient CO_2 concentration increased net ecosystem CO_2 uptake by 50 to 100%, with the greater increase occurring later in the growing season. Net ecosystem CO_2 uptake also increased under CO_2 enrichment in a monospecific stand of the C_4 grass *Spartina patens* (Ait.) Muhl., and in a mixed stand of *Scirpus, Spartina*, and *Distichlis spicata* (L.) Green (C_4), but increases were substantially lower than in the *Scirpus* community. Carbon dioxide enrichment increased net primary productivity in *Scirpus* in both pure stands and in the mixed species stand, but the C_4 species in the mixed species stand were unaffected. Substantial effects of CO_2 enrichment on seed production and plant senescence also occurred in *Scirpus* but not in the C_4 species. These preliminary results confirm that C_3 species in a natural ecosystem probably will benefit more from CO_2 enrichment than will C_4 species.

Tundra ecosystems, because of the small stature of their vegetation and their importance in sequestering C, have also been subjected to direct CO_2 enrichment experiments. These studies generally indicate that doubled CO_2, particularly if accompanied by temperature increases of 4 °C, such as those predicted by many GCM's, will not increase net ecosystem productivity. Billings et al. (1982, 1983) studied the interacting effects of temperature, CO_2 concentration, and depth of water table on net CO_2 exchange from tundra microcosms. They found that increasing the CO_2 concentration from 400 to 800 μmol mol^{-1} increased net C capture, but that decreasing the water table and consequently increasing aerobic respiration greatly increased C loss. Thus doubling of atmospheric CO_2, if accompanied by global warming and lowering of the tundra water table, would cause a net loss in C from the tundra ecosystem.

In subsequent work, Billings et al. (1984) included supplemental N fertilization as a variable. Adding ammonium nitrate greatly increased net CO_2 uptake of the tundra microcosms, even when the temperature was 8 °C. However, increasing CO_2 from 400 to 800 μmol mol^{-1} only slightly increased net CO_2 uptake. Billings et al. (1984) have concluded that if global CO_2 enrichment increases net C gain by the tundra ecosystem, it will be through indirect climatic effects (e.g., warming the soil, increasing decomposition rates, and increasing N availability).

In field studies of a tussock tundra ecosystem, Tissue and Oechel (1987) simulated atmospheric CO_2 doubling and climatic warming in controlled environment enclosures. They concluded that growth of the dominant species (a C_3 sedge, *Eriophorum vaginatum* L.) was limited more by nutrient availability than by ambient atmospheric CO_2 levels. However, CO_2 enrichment to 510 or 680 μmol mol^{-1} substantially increased tiller production. The increase in tillering was reduced by 50%, however, when CO_2 enrichment was accompanied by a 4 °C increase in temperature. Model simulations of arctic tundra ecosystems tend to confirm results of experimental studies: there seems to be no significant effect of CO_2 enrichment on primary productivity (Oechel & Strain, 1985). None of the above studies simulated the natural pattern of diurnal change in the CO_2 concentration. The implications of this are not known.

Model simulations of forest ecosystems are hampered by the lack of information on the long term (>1–2 yr) response of individual trees to CO_2 enrichment (Solomon & West, 1985; Sionit & Kramer, 1986). An early simulation by Botkin et al. (1973), based on the assumption that increased CO_2 would increase annual diameter increments equally in all tree species, predicted that the rates of successional processes would increase and that certain classes of tree species would increase in importance. They predicted, however, that changes in system productivity and biomass would be obscured by these effects and by stochastic population processes. More recent simulation studies based on more sophisticated growth and climatic models indicate that even large increases in the growth rates of individual trees (projected from studies of CO_2 enrichment of seedlings and young trees—the only data now available) will be scaled down considerably when integrated at the forest stand level. Shugart and Emanuel (1985) summarized results from seven model simulations and concluded that a doubling of the growth rate of individual trees would result in only a 35% increase in biomass of mature stands. This down-scaling results from accelerated dieback, differential survival and release of less productive species suppressed under normal conditions, and other complex interactions operating at the ecosystem level.

In unmanaged forests, natural processes—intense interspecific competition, herbivory, disease, fire, dieback, reproduction, succession, and migration—interact in complex ways with climatic variables. Because of severe biological and environmental stresses, very few individual trees reach their maximum potential annual productivity or mature size (Solomon & West, 1985). According to Solomon et al. (1984), in a rapidly changing climate, these stresses change so frequently within the lifetime of individual trees that it is impossible for species better adapted to a particular stress level to replace poorly adapted species. As a result, the productivity of all species declines.

Increasing variability of and rate of change of climatic factors are likely to adversely affect the productivity of unmanaged forests, even if "average" conditions such as increased rainfall and length of growing season appear more favorable (Solomon et al., 1984). Unmanaged forests constitute 97% of the Earth's forested lands. Only 3% are managed plantations (Solomon & West, 1985).

In natural, unmanaged ecosystems, productivity generally is limited by nutrient and water availability (Oechel & Strain, 1985). Therefore, physiological effects of CO_2 enrichment other than enhanced net photosynthesis rates may be the most important determinants of increased productivity in such systems. For example, improved water use efficiency could be critical in ecosystems where the amount and/or timing of precipitation limit plant growth. Furthermore, indirect effects of climate, if they occur, may alter the relative importance of different types of direct physiological effects within the same species and among species (Shugart & Emanuel, 1985).

It is generally accepted that temperature and precipitation control the large-scale distribution of vegetation types, with soil factors superimposed on a more local scale. This assumption has enabled plant geographers to generate maps of hypothetical life zones based on temperature and precipitation patterns. One such system is that of Holdridge (Holdridge, 1947; Holdridge et al., 1971). Emanuel et al. (1985a, b) have combined the Holdridge method with projections of temperature change from a GCM simulation of an Earth with twice ambient atmospheric CO_2 concentration. They projected significant changes in major life zones, as summarized in Table 7–3.

Emanuel et al. (1985b) project a 6% net decrease in the total area of the Earth's life zones that currently support forests. Tropical forest zones would increase in area by 28%, and cool temperate forest zones by 3%. On the other hand, the areas of subtropical forest and warm temperate forest zones would decrease by 22 and 7%, respectively. The area of boreal forest zones would drop by 37%. Arid life zones would increase: grasslands by 25% and deserts by 7%. Life zones supporting tundra would decrease in area by

Table 7-3. Summary of projected changes in life zones with CO_2 doubling.†

Life zone	Current area	Projected change
	$\times\ 10^6\ km^2$	%
Forests—total	75.4	−6
Tropical	19.0	+28
Subtropical	12.0	−22
Warm temperate	15.8	−7
Cool temperate	11.3	+3
Boreal	17.3	−37
Grasslands	22.8	+25
Deserts	26.6	+7
Tundra	4.5	−32
Ice	2.2	−74

† From Emanuel et al., 1985b.

32%. These figures are speculative, but do demonstrate the potentially dramatic consequences of climatic changes associated with increasing atmospheric concentrations of greenhouse gases. Changes in the areal extent of boreal forest and tundra life zones are particularly important, because both of these vegetation types are considered strong sinks for global carbon (Billings, 1987). General circulation models predict disproportionate warming for the high latitudes, which now support these life zones. A decrease in their area of the magnitude projected by Emanuel et al. (1985a, b) could lessen the Earth's ability to "buffer" anthropogenic C release and, during their demise, these vegetation types would become additional sources of atmospheric CO_2 (Billings, 1987).

The consensus among ecologists is that the projected climatic changes associated with increases in the concentration of the greenhouse gases will generally be disruptive of natural or unmanaged ecosystems (Solomon et al., 1984; Strain & Armentano, 1982; Strain, 1987). As noted previously, the projected rate of temperature increase expected to result from doubled atmospheric CO_2 concentration is unprecedented in the earth's history. The expected change in mean global temperature, equivalent in magnitude to those that have extended over millenia in the past, will be compressed into a few centuries or less. During past episodes of gradual climatic change, latitudinal and elevational migrations allowed many species of plants and animals to survive, but major species extinctions also occurred (Billings, 1965). Rates of migration, particularly for perennial plants, are too slow by several orders of magnitude to cope with the rapid changes projected for the next several centuries (Warrick et al., 1986b; Shugart et al., 1986). Thus, it is likely that the species that do survive will be those with wide environmental tolerance ranges, typically "weedy" plants (Baker, 1974). Rare species or those restricted to odd local climates are especially likely to become extinct.

There is increasing concern over the effects of projected climate change on the biological reserves that have been established to conserve rare species of plants and animals (Peters & Darling, 1985). These reserves were located assuming a stable climate. If the climate changes, the environmental requirements of the protected species will not be met. Migration is unlikely, because bioreserves often are surrounded by developed or agricultural areas. Therefore, some protected species will probably be lost.

EFFECTS OF INCREASING CARBON DIOXIDE AND CLIMATIC CHANGE ON MANAGED ECOSYSTEMS

The climatic consequences of increasing concentrations of greenhouse gases may be less severe for intensively managed ecosystems such as annual crops, pastures, and short-rotation commercial forests (Wittwer, 1985; Waggoner, 1983). In these managed ecosystems, adapted species or cultivars are selected, and environmental or biological stresses may be relieved by fertilization, irrigation, pest control, and adjustment of planting dates. Rosenberg (1981, 1982) and Waggoner (1983) emphasize the resiliency of modern agricultural production systems and the demonstrated ability of agriculture to adapt to short-term variations in climate experienced since the late 19th century. However, there are still limits to the abilities of technology to deal with extreme climatic events, even in high-input temperate agricultural systems (Oram, 1985). Furthermore, in the less technologically advanced agricultural production systems still prevalent in many areas of the world, the management options are very limited. These systems may be expected to be much more vulnerable to the adverse effects of a rapidly changing climate (Warrick et al., 1986a, b).

Several projections have been made of climatic effects of an equivalent CO_2 doubling on specific modern agricultural production systems. For example, Blasing and Solomon (1984) evaluated the potential effects on the North American Corn Belt of the 3 °C increase in winter and summer temperature predicted for the region by Manabe and Stouffer (1980) as a consequence of CO_2 doubling. Blasing and Solomon (1984) incorporated potential changes in precipitation and moisture availability in their projection, but did not include possible improvements in agricultural technology or favorable direct effects of enriched CO_2 on corn growth. Based on changes in degree days and precipitation, they predicted a substantial northward and northeastward shift of the Corn Belt zones, and an increase in the need for irrigation within the boundaries of the present Corn Belt as well as in the new Corn Belt zone.

Rosenzweig (1985) analyzed the possible effects of a 4.2 °C mean global warming on wheat (*Triticum aestivum* L.) production areas in North America. Consideration of growing season length, accumulated degree days, mean minimum January temperature, and annual precipitation in the hypothetical new climate yielded predictions of significant shifts in wheat-producing zones. For example, warmer temperatures are expected to extend the winter wheat belt into Canada. In Mexico, warmer temperatures would probably increase stress levels, but would not force major shifts in production areas. Within the USA, shifts in the localization of winter vs. spring and soft vs. hard wheat types were projected, but the major wheat-growing zones probably would not change markedly. An important contribution of the Rosenzweig study was the identification of *areas of vulnerability*. Thus, wheat production regions in Mexico would be subject to greater likelihood of high-temperature stress, whereas those in the U.S. southern Great Plains would be subject to increased water stress.

Palutikof et al. (1984) assessed potential effects of CO_2-induced global warming on yields of wheat, barley (*Hordeum vulgare* L.), oat (*Avena sativa* L.), hay, turnip (*Brassica rapa*), and swede (*Brassica napus*) in England and Wales. They based their analysis on scenarios constructed from temperature and precipitation data from naturally occurring warm and cool periods during this century. Substantial differences in seasonal distribution of precipitation occurred during the warm and cold periods, in addition to the obvious differences in accumulated degree days and length of the growing season. After combining the projected climatic change with crop-weather regression models, Palutikof et al.(1984) concluded that wheat yields were unlikely to be affected, but that yields of barley, oat, hay, and the root crops turnip and swede would be expected to decrease somewhat.

None of the above studies incorporated potential improvements in agricultural technology and management. As pointed out by Waggoner (1983), Wittwer (1985), Rosenberg (1982), and others, these factors may ameliorate some possible negative effects of climatic warming; this is demonstrated in a more recent simulation study. Wilks (1988) incorporated management options such as planting date and cultivar selection into a simulation of the effects of CO_2 doubling on corn and winter and spring wheat production in three areas of the USA. He used process-oriented crop yield models to investigate the effects of climatic changes derived from a GCM simulation. The results indicate that choosing corn varieties and planting dates adapted to the projected high-CO_2 climate can reduce the deleterious effects of projected increases in evapotranspiration and water stress during critical growth periods. Elevated temperatures during late growth stages are more likely to adversely affect wheat production, but again changes in planting dates and increased rates of vegetative growth associated with higher winter temperatures can reduce deleterious effects. Wilks' study illustrates the complexities of the stresses and management options that must be considered in projecting future crop yields in the warmer high-CO_2 world.

Wilks' (1988) simulations did not incorporate the largely beneficial direct physiological effects of CO_2 enrichment, although he emphasized the desirability of doing so. If these physiological effects, which frequently have been demonstrated in controlled-environment CO_2 enrichment studies, translate to crops grown in monoculture in the field, some of the deleterious effects associated with climatic change may be reduced (Rosenberg, 1981, 1982; Waggoner, 1983; Wittwer, 1985). Recent experimental work indicates that some of the beneficial effects of CO_2 enrichment on crop biomass production previously observed in plants grown singly in pots will extrapolate to plants grown in rows in the field. Rogers et al. (1986) reported biomass production of individual soybean plants grown in rows to increase significantly with CO_2 enrichment to approximately twice ambient levels in open-top chambers. However, because of a decreased harvest index in well-watered conditions, pod yield was not increased. Jones et al. (1984) reported substantial biomass and seed yield increases in field-grown soybean exposed to 600 μmol mol^{-1} CO_2 in controlled-environment canopy enclosures. Kimball et al. (1985) reported significant increases in both biomass and economic yield in field-grown cotton (*Gossypium hirsutum* L.) exposed to doubled CO_2 in open-top chambers. Significant biomass and yield increases also occurred in cotton exposed to free air CO_2 enrichment of approximately 200 μmol mol^{-1} above ambient (Kimball et al., 1986). Results of both Rogers et al. (1986) and Kimball et al. (1986) further indicate that CO_2 enrichment increases biomass production. These studies are an important step in demonstrating that the direct physiological effects of global atmospheric CO_2 enrichment can increase yield in crops grown in monoculture under field conditions.

However, few crops in agricultural fields are grown in true monoculture. In agroecosystems, weeds compete directly with crops for the limited resources upon which plant growth depends. In the USA, weeds cause annual yield losses estimated at 10% of potential production, valued at $12 billion. These losses occur in spite of $6.2 billion spent annually on weed control, so the total annual loss due to weeds exceeds $18 billion (Shaw, 1982; Waggoner, 1983; Wittwer, 1985). Other crop pests, insects, and plant diseases also significantly reduce the quantity and the quality of the yield of many crops (Waggoner, 1983).

IMPLICATIONS OF INCREASING CARBON DIOXIDE AND CLIMATIC CHANGE FOR WEED AND PEST MANAGEMENT

A major obstacle to the extrapolation of the results of controlled-environment CO_2 enrichment studies to agricultural production systems is lack of knowledge of the potential effects of increased CO_2 and climatic change on agricultural pests and their control (NRC, 1976; Waggoner, 1983). For example, if increased CO_2 acts as a fertilizer, the growth of weeds as well as crops will be stimulated, perhaps to the detriment of the crops. Field experiments with conventional fertilizers have indicated increased weed competition can occur in response to supplemental nutrients (Vengris et al., 1955; Gruenhagen & Nalewaja, 1969). The presence of competing weeds can eliminate the yield-enhancing effects of fertilization (Fabricius & Nalewaja, 1968; Patterson, 1985).

The widely reported differential responses of C_3 and C_4 plants to CO_2 enrichment are particularly relevant to weed/crop competition in future agroecosystems in a high CO_2 world. Most crops are C_3 plants. Of the "15 crops that feed the world" (Harlan, 1975), 12 have the C_3 pathway. On the other hand, 14 of the 18 "world's worst weeds" (Holm et al., 1977) are C_4 plants (Patterson, 1985). Many major weeds of the few C_4 crops are C_3 plants. For example, 19 of the 38 major weeds of corn in the USA are C_3 plants.

It is in agroecosystems that competition among C_3 and C_4 plants is most intense. In natural ecosystems such as shortgrass prairie, the growth of C_3 and C_4 species may be temporally separated, reducing the effective degree of competition. However, in agroecosystems the growth of weeds and crops is synchronized by soil preparation and other management processes that create regular, periodic disturbance. Competition, therefore, is maximized among plants that do not even occur together in natural ecosystems.

As discussed previously, the direct physiological effects of CO_2 enrichment are likely to benefit C_3 plants more than their C_4 competitors. This has been confirmed in several studies (Carter & Peterson, 1983; Bazzaz & Carlson, 1984; Patterson et al., 1984; Zangerl & Bazzaz, 1984; Wray & Strain, 1987a). However, interactions with other environmental factors may override the direct influence of CO_2 so that the outcome of competition between C_3 and C_4 plants cannot always be predicted based only on their response to CO_2 (Bazzaz et al., 1989).

Indeed, it seems likely that the indirect climatic effects of increasing CO_2 will benefit C_4 plants more than their C_3 competitors, particularly if global warming is accompanied by increasing regional aridity and frequency of water stress, as predicted for important crop-producing regions (Warrick et al., 1986a). Chu et al. (1978), Pearcy et al. (1981), Patterson (1982), and Flint and Patterson (1983) all have reported that increased temperatures favor C_4 species in competition with C_3 species. Ozturk et al. (1981) reported that limited soil water availability favored C_4 species in competition with C_3 species.

The foregoing discussion should not imply that increased CO_2 and climate change will affect competition only between weed and crop species differing in photosynthetic pathways. Within groups of weed species sharing the same photosynthetic pathway there are significant differences in growth response to CO_2 enrichment, particularly in competitive situations in which other environmental factors are varied (Bazzaz & Carlson, 1984; Zangerl & Bazzaz, 1984). Although our knowledge of potential weed-crop interactions for species sharing the same pathway is limited, these studies suggest that CO_2 enrichment may well alter the competitive balance between C_3 crops and C_3 weeds, and between C_4 crops and weeds.

Both the direct and indirect effects of increased CO_2 may be expected to affect the geographical distribution of weeds and other crop pests (Waggoner, 1983). Increased CO_2 may improve the ability of plants sensitive to chilling to grow at suboptimum temperatures (Sionit et al., 1981b; Sasek & Strain, 1988). This has important implications for weed/crop competition, because many of the most aggressive agricultural weeds in temperate crops are of tropical or subtropical origin, and their distribution currently is limited by unfavorably low temperatures. For example, Patterson and Flint (1979) reported that short-term chilling events inhibited the growth of itchgrass [*Rottboellia exaltata* (L.) f.] more than that of adapted varieties of corn and soybean. Thus itchgrass is unlikely to be competitive outside the South under current climatic conditions. The growth of itchgrass is greatly reduced by mean temperatures below 29/23 °C (Patterson et al., 1979). Its impact as a weed in corn and soybean would be expected to increase with the substantial CO_2-induced temperature increases projected for the U.S. Corn Belt. Similarly, witchweed [*Striga asiatica* (L.) Kuntze], a major parasite of corn in tropical and subtropical regions, currently is confined in this country to the coastal plain of the Carolinas (Patterson et al., 1982). Increasing temperature could allow it to invade the Corn Belt with potentially devastating results.

Sasek and Strain (1989) discussed the potential effects of increased CO_2 and climate change on the distribution of two exotic weeds of forest ecosystems. The current distribution of both Japanese honeysuckle (*Lonicera japonica* Thunb.) and kudzu [*Pueraria lobata* (Willd.) Ohwi] apparently is limited by low winter temperatures. If these temperatures increase by 3 °C as projected, the northern limits of honeysuckle and kudzu could be extended by several hundred kilometers. Increased CO_2 also enhances the growth of these species at suboptimum temperatures, and this effect would contribute to northward expansion. The western limits of kudzu and honeysuckle probably are controlled by moisture availability. These limits might decrease in response to CO_2-induced warming, whereas the water use efficiency of both species could increase in direct response to CO_2 enrichment. Thus, the western limits of distribution of these species are unlikely to change, according to Sasek and Strain (1989). This exemplary study has far-reaching implications for the distribution of major agricultural weeds.

In addition to affecting the potential competitiveness and geographical distribution of agricultural weeds, increasing CO_2 and climate change probably will influence human efforts to control weeds. Environmental factors such as temperature, soil moisture, and humidity are well known to influence the effectiveness of herbicides on targeted weeds (Muzik, 1976; Hatzios & Penner, 1982). Variations in these environmental factors also may influence crop injury by herbicides. The implications of these environmental effects have not been explored in the context of increasing CO_2 and climate change. It seems likely, however, that projected changes in temperature and moisture availability may affect both the uptake and metabolism of herbicides by both weeds and crops (Hatzios & Penner, 1982). The antitranspirant effects of high CO_2 might reduce uptake of soil-applied herbicides, but the significance of these effects under field conditions is unknown. Changes in leaf anatomy and surface characteristics induced by high CO_2 could affect uptake of herbicides applied to foliage. Increased rhizome and tuber growth in difficult-to-control perennial weeds is an likely consequence of enhanced photosynthate production under high CO_2. This should increase the resistance of such weeds to chemical control.

Natural and implemented biological control of weeds and other crop pests also probably will be affected by increasing CO_2 and climatic change (Andres, 1977; Norris, 1982). Synchrony between the growth, development, and reproduction of biocontrol agents and of their target pests is essential to successful control (Messenger et al., 1976). Such synchrony may well be disrupted by climatic change.

All of these projections are only speculative, but such possibilities must be taken into account when the results of controlled environment studies of CO_2 enrichment are extrapolated to agricultural production systems.

CONCLUSIONS

Continuing increases in the CO_2 concentration of the Earth's atmosphere will directly affect the physiological processes and growth of plants. Indirect climatic effects, including global warming and changes in precipitation patterns and the frequency of weather extremes, probably will have greater impact than direct effects of CO_2 on physiological processes. Increasing atmospheric CO_2 to twice ambient levels generally will increase plant growth, although the magnitude of growth stimulation will vary greatly with species, photosynthetic pathway, growth stage, and water and nutrient status. Differential growth responses to both CO_2 concentration and climatic change will affect future competitive ability and fitness of plants in both natural and managed ecosystems. The relative importance of various weed species in agroecosystems may change, but selection of adapted crop varieties and management methods probably will minimize negative impacts. In natural ecosystems, species extinctions probably will increase because migration and adaptation through natural selection may be inadequate to accommodate the rapid climatic changes involved. Weeds and other plants with broad ecological amplitudes are likely to prosper at the expense of endemic species or species already in marginal habitats.

ACKNOWLEDGMENTS

We thank Ms. L.K. Sentelle for typing the manuscript and Drs. L.H. Allen, Jr., J.D. Hesketh, B.A. Kimball, N.J. Rosenberg, and T.W. Sasek for reviewing the manuscript.

REFERENCES

Andres, L.A. 1977. The biological control of weeds. p. 153–176. *In* J.D. Fryer and S. Matsunaka (ed.) Integrated control of weeds. Univ. of Tokyo Press, Tokyo.

Baker, H.G. 1974. The evolution of weeds. Ann. Rev. Ecol. Syst. 5:1–24.

Baker, J.T., L.H. Allen, Jr., K.J. Boote, P. Jones, and J.W. Jones. 1989. Response of soybean to air temperature and carbon dioxide concentration. Crop Sci. 29:98–105.

Barlow, E.W.R., and J. Conroy. 1988. Influence of elevated atmospheric carbon dioxide on the productivity of Australian forestry plantation. p. 520–533. *In* G.I. Pearman (ed.) Greenhouse: Planning for climate change. E.J. Brill, Leiden.

Bazzaz, F.A., and R.W. Carlson. 1984. The response of plants to elevated CO_2. I. Competition among an assemblage of annuals at two levels of soil moisture. Oecologia 62:196–198.

Bazzaz, F.A., and K. Garbutt. 1988. The response of annuals in competitive neighborhoods: Effects of elevated CO_2. Ecology 69:937–946.

Bazzaz, F.A., K. Garbutt, E.G. Reekie, and W.E. Williams. 1989. Using growth analysis to interpret competition between a C_3 and a C_4 annual under ambient and elevated CO_2. Oecologia 79:223–235.

Beer, T., A.M. Gill, and P.H.R. Moore. 1988. Australian bushfire danger under changing climatic regimes. p. 421–427. *In* G.I. Pearman (ed.) Greenhouse: Planning for climate change. E.J. Brill, Leiden.

Begon, M., J.L. Harper, and C.R. Townsend. 1986. Ecology: Individuals, populations and communities. Sinauer Associates, Sunderland, MA.

Billings, W.D. 1965. Plants and the ecosystem. Wadsworth Publishing Co., Belmont, CA.

Billings, W.D. 1987. Carbon balance of Alaskan tundra and taiga ecosystems: Past, present and future. Quat. Sci. Rev. 6:165–177.

Billings, W.D., J.O. Luken, D.A. Mortensen, and K. Peterson. 1982. Arctic tundra: A source or sink for atmospheric carbon dioxide in a changing environment? Oecologia 53:7–11.

Billings, W.D., J.O. Luken, D.A. Mortensen, and K. Peterson. 1983. Increasing atmospheric carbon dioxide: Possible effects on arctic tundra. Oecologia 58:286–289.

Billings, W.D., K.M. Peterson, J.O. Luken, and D.A. Mortensen. 1984. Interaction of increasing atmospheric carbon dioxide and soil nitrogen on the carbon balance of tundra microcosms. Oecologia 65:26–29.

Blasing, T.J., and A.M. Solomon. 1984. Response of the North American corn belt to climatic warming. Prog. Biometeorol. 3:311–321.

Bolin, B., J. Jager, and B.R. Doos. 1986. The greenhouse effect, climatic change, and ecosystems: A synthesis of present knowledge. p. 1–32. *In* B. Bolin et al. (ed.) SCOPE 29. The greenhouse effect, climatic change, and ecosystems. John Wiley & Sons, New York.

Bolle, H.-J., W. Seiler, and B. Bolin. 1986. Other greenhouse gases and aerosols: Assessing their role for atmospheric radiative transfer. p. 93–156. *In* B. Bolin et al. (ed.) SCOPE 29. The greenhouse effect, climatic change, and ecosystems. John Wiley & Sons, New York.

Botkin, D.B., J.F. Janak, and J.R. Wallis. 1973. Estimating the effects of carbon fertilization on forest composition by ecosystem simulation. p. 328–344. *In* G.M. Woodwell and E.V. Pecan (ed.) Carbon and the biosphere. A.E.C. Symp. Ser. 30. NTIS, Springfield, VA.

Carlson, R.W., and F.A. Bazzaz. 1980. The effects of elevated CO_2 concentrations on growth, photosynthesis, transpiration and water use efficiency of plants. p. 610–622. *In* J.J. Singh and A. Deepak (ed.) Environmental and climatic impact of coal utilization. Academic Press, New York.

Carlson, R.W., and F.A. Bazzaz. 1982. Photosynthetic and growth response to fumigation with SO_2 at elevated CO_2 for C_3 and C_4 plants. Oecologia 54:50–54.

Carter, D.R., and K.M. Peterson. 1983. Effects of a CO_2 enriched atmosphere on the growth and competitive interaction of a C_3 and a C_4 grass. Oecologia 58:188–193.

Chu, C., P.M. Lundford, J.L. Ozbun, and R.D. Sweet. 1978. Effects of temperature and competition on the establishment and growth of redroot pigweed and common lambsquarters. Crop Sci. 18:308–310.

Coleman, J.M. 1988. Climatic warming and increased summer aridity in Florida, U.S.A. Clim. Change 12:165–178.

Cure, J.D. 1985. Carbon dioxide doubling responses: A crop survey. p. 99–116. *In* B.R. Strain and J.D. Cure (ed.) Direct effects of increasing carbon dioxide on vegetation. DOE/ER-0238. U.S. Dep. of Energy, Washington, DC. Available from NTIS, Springfield, VA.

Dickinson, R.E. 1986. How will climate change? The climate system and modelling of future climate. p. 207–270. *In* B. Bolin et al. (ed.) SCOPE 29. The greenhouse effect, Climatic change, and ecosystems. John Wiley & Sons, New York.

Downton, W.J.S., W.J.R. Grant, and B.R. Loveys. 1987. Carbon dioxide enrichment increases yield of Valencia orange. Aust. J. Plant Physiol. 14:493–501.

Drake, B.G., P.S. Curtis, W.J. Arp, P.W. Leadley, J. Johnson, and D. Whigham. 1988. Effects of elevated CO_2 on Chesapeake Bay wetlands. III. Ecosystem and whole plant responses in the first year of exposure, April–November 1987. Response of vegetation to carbon dioxide, report 044. U.S. Dep. of Energy, Washington, DC.

Emanuel, W.R., H.H. Shugart, and M.P. Stevenson. 1985a. Climatic change and the broad-scale distribution of terrestrial ecosystem complexes. Clim. Change 7:29–43.

Emanuel, W.R., H.H. Shugart, and M.P. Stevenson. 1985b. Response to comment: Climatic change and the broad-scale distribution of terrestrial ecosystem complexes. Clim. Change 7:457–460.

Fabricius, L.J., and J.D. Nalewaja. 1968. Competition between wheat and wild buckwheat. Weed Sci. 16:204–208.

Flint, E.P., and D.T. Patterson. 1983. Interference and temperature effects on growth in soybean (*Glycine max*) and associated C_3 and C_4 weeds. Weed Sci. 31:193–199.

Garbutt, K., and F.A. Bazzaz. 1984. The effects of elevated CO_2 on plants. III. Flower, fruit and seed production and abortion. New Phytol. 98:433–446.

Gifford, R.M. 1988. Direct effects of higher carbon dioxide concentrations on vegetation. p. 506–519. *In* G.I. Pearman (ed.) Greenhouse: Planning for climate change. E.J. Brill, Leiden.

Gifford, R.M., H. Lambers, and J.I.L. Morison. 1985. Respiration of crop species under CO_2 enrichment. Physiol. Plant. 63:351–356.

Goudriaan, J., and H.E. de Ruiter. 1983. Plant growth in response to CO_2 enrichment, at two levels of nitrogen and phosphorus supply. 1. Dry matter, leaf area and development. Neth. J. Agric. 31:157–169.

Graetz, R.D. 1988. Climate change and terrestrial vegetation. p. 126–129. *In* Global Change: Proc. of the Elizabeth and Frederick White Research Conf., Canberra. 24–27 February. Australian Academy of Science, Canberra.

Grime, J.P. 1979. Plant strategies and vegetation processes. John Wiley & Sons, New York.

Gruenhagen, R.D., and J.D. Nalewaja. 1969. Competition between flax and wild buckwheat. Weed Sci. 17:380–384.

Harlan, J.R. 1975. Crops and man. ASA and CSSA, Madison, WI.

Hatzios, K.K., and D. Penner. 1982. Metabolism of herbicides in higher plants. CEPCO Div., Burgess Publishing, Edina, MN.

Hofstra, G., and J.D. Hesketh. 1975. The effects of temperature and CO_2 enrichment on photosynthesis in soybean. p. 71–80. *In* R. Marcelle (ed.) Environmental and biological control of photosynthesis. Junk, The Hague, Netherlands.

Holdridge, L.R. 1947. Determination of world plant formations from simple climatic data. Science (Washington, DC) 105:367–368.

Holdridge, L.R., W.C. Grenke, W.H. Hatheway, T. Liang, and J.A. Tosi, Jr. 1971. Forest environments in tropical life zones: A pilot study. Pergamon Press, Elmsford, NY.

Holm, L.G., D.L. Plucknett, J.V. Pancho, and J.P. Herberger. 1977. The world's worst weeds: Distribution and biology. Univ. of Hawaii Press, Honolulu.

Houghton, R.A., W.H. Schlesinger, S. Brown, and J.F. Richards. 1985. Carbon dioxide exchange between the atmosphere and terrestrial ecosystems. p. 113–140. *In* J.R. Trabalka (ed.) Atmospheric carbon dioxide and the global carbon cycle. DOE/ER-0239. U.S. Dep. of Energy, Washington, DC. Available from NTIS, Springfield, VA.

Idso, S.B., B.A. Kimball, M.G. Anderson, and J.R. Mauney. 1987. Effects of atmospheric CO_2 enrichment on plant growth: The interactive role of air temperature. Agric. Ecosyst. Environ. 20:1–10.

Jones, P., L.H. Allen, Jr., J.W. Jones, K.J. Boote, and W.J. Campbell. 1984. Soybean canopy growth, photosynthesis, and transpiration responses to whole-season carbon dioxide enrichment. Agron. J. 76:633–637.

Kimball, B.A. 1983. Carbon dioxide and agricultural yield: An assemblage and analysis of 430 prior observations. Agron. J. 75:779–788.

Kimball, B.A. 1986a. CO_2 stimulation of growth and yield under environmental restraints. p. 53–67. *In* H.Z. Enoch and B.A. Kimball (ed.) Carbon dioxide enrichment of greenhouse crops. Vol. 2. Physiology, yield, and economics. CRC Press, Boca Raton, FL.

Kimball, B.A. 1986b. Influence of elevated CO_2 on crop yield. p. 105–115. *In* H.Z. Enoch and B.A. Kimball (ed.) Carbon dioxide enrichment of greenhouse crops. Vol. 2. Physiology, yield, and economics. CRC Press, Boca Raton, FL.

Kimball, B.A., J.R. Mauney, G. Guinn, F.S. Nakayama, S.B. Idso, J.W. Radin, D.L. Hendrix, G.D. Butler, T.I. Zarembinski, and P.E. Nixon II. 1985. Effects of increasing atmospheric CO_2 on the yield and water use of crops. Response of vegetation to carbon dioxide. Rep. 027. USDA and U.S. Dep. of Energy, Washington, DC.

Kimball, B.A., J.R. Mauney, J.W. Radin, F.S. Nakayama, S.B. Idso, D.L. Hendrix, D.H. Akey, S.G. Allen, M.G. Anderson, and W. Hartung. 1986. Effects of increasing atmospheric CO_2 on the growth, water relations, and physiology of plants grown under optimal and limiting levels of water and nitrogen. Response of vegetation to carbon dioxide. Rep. 039. USDA and U.S. Dep. of Energy, Washington, DC.

Kramer, P.J. 1981. Carbon dioxide concentration, photosynthesis, and dry matter production. BioScience 31:29–33.

Lamborg, M.R., R.W.F. Hardy, and E.A. Paul. 1983. Microbial effects. p. 131–176. *In* E.R. Lemon (ed.) CO_2 and plants: The response of plants to rising levels of atmospheric carbon dioxide. Westview Press, Boulder, CO.

Larigauderie, A., D.W. Hilbert, and W.C. Oechel. 1988. Effect of CO_2 enrichment and nitrogen availability on resource acquisition and resource allocation processes in a grass, *Bromus mollis*. Oecologia 77:544–549.

Lincoln, D.E., D. Couvet, and N. Sionit. 1986. Response of an insect herbivore to host plants grown in carbon dioxide enriched atmospheres. Oecologia 69:556–560.

Lincoln, D.E., N. Sionit, and B.R. Strain. 1984. Growth and feeding response of *Pseudoplusia includens* (Lepidoptera: Noctuidae) to host plants grown in controlled carbon dioxide atmospheres. Environ. Ent. 13:1527–1530.

Luther, F.M., and R.D. Cess. 1985. Review of the recent carbon dioxide-climate controversy. p. 321–335. *In* M.C. McCracken and F.M. Luther (ed.) Projecting the climatic effects of increasing carbon dioxide. DOE/ER-0237. U.S. Dep. of Energy, Washington, DC. Available from NTIS, Springfield, VA.

Manabe, S., and R.J. Stouffer. 1980. Sensitivity of a global climate model to an increase of CO_2 concentration in the atmosphere. J. Geophys. Res. 85:5529–5554.

Manabe, S., and T. Wetherald. 1986. Reduction in summer soil wetness induced by an increase in atmospheric carbon dioxide. Science (Washington, DC) 232:626–628.

Marks, S., and B.R. Strain. 1989. Effects of drought and CO_2 enrichment on competition between two old-field perennials. New Phytol. 111:181–186.

Mearns, L.O., R.W. Katz, and S.H. Schneider. 1984. Extreme high temperature events: Changes in their probabilities with changes in mean temperature. J. Clim. Appl. Meteorol. 23:1601–1613.

Messenger, P.S., E. Biliotti, and R. van den Bosch. 1976. The importance of natural enemies in integrated control. p. 543–563. *In* C.B. Huffaker and P.S. Messenger (ed.) Theory and practice of biological control. Academic Press, New York.

Muzik, T.J. 1976. Influence of environmental factors on toxicity to plants. p. 204–247. *In* L.L. Audus (ed.) Herbicides: Physiology, biochemistry, ecology. 2nd ed. Vol. 2. Academic Press, New York.

National Research Council. 1976. Pest management. p. 107–130. *In* Climate & food: Climatic fluctuation and U.S. agricultural production. Nat. Acad. Sci., Washington, DC.

Nijs, I., I. Impens, and T. Behaege. 1988. Effects of rising atmospheric carbon dioxide concentration on gas exchange and growth of perennial ryegrass. Photosynthetica 22:44–50.

Norris, R.F. 1982. Interactions between weeds and other pests in the agro-ecosystem. p. 343–406. *In* J.L. Hatfield and I.J. Thomason (ed.) Biometeorology in integrated pest management. Academic Press, New York.

Oberbauer, S.F., N. Sionit, S.J. Hastings, and W.C. Oechel. 1986. Effects of CO_2 enrichment and nutrition on growth, photosynthesis and nutrient concentration of Alaska tundra plant species. Can. J. Bot. 64:2993–2998.

Oberbauer, S.F., B.R. Strain, and N. Fetcher. 1985. Effect of CO_2-enrichment on seedling physiology and growth of two tropical tree species. Physiol. Plant. 65:352–356.

Oechel, W.C., and B.R. Strain. 1985. Native species responses to increased carbon dioxide concentration. p. 117–154. *In* B.R. Strain and J.D. Cure (ed.) Direct effects of increasing carbon dioxide on vegetation. DOE/ER-0238. U.S. Dep. of Energy, Washington, DC. Available from NTIS, Springfield, VA.

Oram, P.A. 1985. Sensitivity of agricultural production to climatic change. Clim. Change 7:129–152.

Overdieck, D., D. Bossemeyer, and H. Lieth. 1984. Long-term effects of an increased CO_2 concentration level on terrestrial plants in model-ecosystem I. Phytomass production and competition of *Trifolium repens* L. and *Lolium perenne* L. Prog. Biometeorol. 3:344–352.

Overdieck, D.,C. Reid, and B.R. Strain. 1988. The effects of preindustrial and future CO_2 concentrations on growth, dry matter production and the C/N relationship in plants at low nutrient supply: *Vigna unguiculata* (cowpea), *Abelmoschus esculentus* (okra) and *Raphanus sativus* (radish). Angew. Bot. 62:119–134.

Overdieck, D., and F. Reining. 1986. Effect of atmospheric CO_2 enrichment on perennial ryegrass (*Lolium perenne* L.) and white clover (*Trifolium repens* L.) competing in managed model-ecosystems. Acta Oecol. Oecol. Plant. 7:357–366.

Ozturk, M., H. Rehder, and H. Ziegler. 1981. Biomass production of C_3- and C_4-plant species in pure and mixed culture with different water supply. Oecologia 50:73–81.

Palutikof, J.P., T.M.L. Wigley, and G. Farmer. 1984. The impact of CO_2 induced climate change on crop-yields in England and Wales. Prog. Biometeorol. 3:320–334.

Parry, M.L., and T.R. Carter. 1985. The effect of climatic variations on agricultural risk. Clim. Change 7:95–110.

Patterson, D.T. 1982. Effects of light and temperature on weed/crop growth and competition. p. 407–420. *In* J.L. Hatfield and I.J. Thomason (ed.) Biometeorology in integrated pest management. Academic Press, New York.

Patterson, D.T. 1985. Comparative ecophysiology of weeds and crops. p. 101–129. *In* S.O. Duke (ed.) Weed physiology. Vol. 1. Reproduction and ecophysiology. CRC Press, Boca Raton, FL.

Patterson, D.T. 1986. Responses of soybean (*Glycine max*) and three C_4 grass weeds to CO_2 enrichment during drought. Weed Sci. 34:203–210.

Patterson, D.T., and E.P. Flint. 1979. Effects of simulated field temperatures and chilling on itchgrass (*Rottboellia exaltata*), corn (*Zea mays*), and soybean (*Glycine max*). Weed Sci. 27:645–650.

Patterson, D.T., and E.P. Flint. 1980. Potential effects of global atmospheric CO_2 enrichment on the growth and competitiveness of C_3 and C_4 weed and crop plants. Weed Sci. 28:71–75.

Patterson, D.T., and E.P. Flint. 1982. Interacting effects of CO_2 and nutrient concentration. Weed Sci. 30:389–394.

Patterson, D.T., E.P. Flint, and J.L. Beyers. 1984. Effects of CO_2 enrichment on competition between a C_4 weed and a C_3 crop. Weed Sci. 32:101–105.

Patterson, D.T., M.T. Highsmith, and E.P. Flint. 1988. Effects of temperature and CO_2 concentration on the growth of cotton (*Gossypium hirsutum*), spurred anoda (*Anoda cristata*), and velvetleaf (*Abutilon theophrasti*). Weed Sci. 36:751–757.

Patterson, D.T., C.R. Meyer, E.P. Flint, and P.C. Quimby, Jr. 1979. Temperature responses and potential distribution of itchgrass (*Rottboellia exaltata*) in the United States. Weed Sci. 27:77–82.

Patterson, D.T., R.L. Musser, E.P. Flint, and R.E. Eplee. 1982. Temperature responses and potential for spread of witchweed (*Striga lutea*) in the United States. Weed Sci. 30:87–93.

Pearcy, R.W., N. Tumosa, and K. Williams. 1981. Relationships between growth, photosynthesis and competitive interactions for a C_3 and a C_4 plant. Oecologia 48:371–376.

Peters, R.L., and J.D.S. Darling. 1985. The greenhouse effect and nature reserves. BioScience 35:707–717.

Pittock, A.B. 1988. Regional climatic change and its impacts. p. 98–109. *In* Global change: Proc. of the Elizabeth and Frederick White Research Conf., Canberra. 24–27 February. Australian Academy of Science, Canberra.

Poorter, H., S. Pot, and H. Lambers. 1988. The effect of an elevated atmospheric CO_2 concentration on growth, photosynthesis and respiration of *Plantago major*. Physiol. Plant. 73:553–559.

Potvin, C., and B.R. Strain. 1985. Effects of CO_2 enrichment and temperature on growth in two C_4 weeds, *Echinochloa crus-galli* and *Eleusine indica*. Can. J. Bot. 63:1495–1499.

Reekie, E.G., and F.A. Bazzaz. 1989. Competition and patterns of resource use among seedlings of five tropical trees grown at ambient and elevated CO_2. Oceologia 79:212–222.

Riechers, G.H., and B.R. Strain. 1988. Growth of blue grama (*Bouteloua gracilis*) in response to atmospheric CO_2 enrichment. Can. J.Bot. 66:1570–1573.

Rogers, H.H., G.E. Bingham, J.D. Cure, J.M. Smith, and K.A. Surano. 1983. Responses of selected plant species to elevated carbon dioxide in the field. J. Environ. Qual. 12:569–574.

Rogers, H.H., J.D. Cure, and J.M. Smith. 1986. Soybean growth and yield response to elevated carbon dioxide. Agric. Ecosyst. Environ. 16:113–128.

Rosenberg, N.J. 1981. The increasing CO_2 concentration in the atmosphere and its implication on agricultural productivity. I. Effects on photosynthesis, transpiration and water use efficiency. Clim. Change 3:265–279.

Rosenberg, N.J. 1982. The increasing CO_2 concentration in the atmosphere and its implication on agricultural productivity. II. Effects through CO_2 induced climatic change. Clim. Change 4:239–254.

Rosenzweig, C. 1985. Potential CO_2 induced climate effects on North American wheat-producing regions. Clim. Change 7:367–389.

Sage, R.F., and R.D. Sharkey. 1987. The effect of temperature on the occurrence of O_2 and CO_2 insensitive photosynthesis in field grown plants. Plant Physiol. 78:619–622.

Sasek, T.W., and B.R. Strain. 1988. Effects of carbon dioxide enrichment on the growth and morphology of kudzu (*Pueraria lobata*). Weed Sci. 36:28–36.

Sasek, T.W., and B.R. Strain. 1989. Implications of atmospheric CO_2 enrichment and climate change for the geographical distribution of two introduced vines in the U.S.A. Clim. Change (in press).

Schlesinger, M.E., and J.F.B. Mitchell. 1985. Model projections of the equilibrium climatic response to increased carbon dioxide. p. 81–147. *In* M.C. McCracken and F.M. Luther (ed.) Projecting the climatic effects of increasing carbon dioxide. DOE/ER-0237. U.S. Dep. of Energy, Washington, DC. Available from NTIS, Springfield, VA.

Shaw, W.C. 1982. Integrated weed management systems technology for pest management. Weed Sci. 30 (Suppl. 1):2–12.

Shugart, H.H., M.Y. Antonovsky, P.G. Jarvis, and A.P. Sandford. 1986. CO_2, climatic change and forest ecosystems. p. 475–521. *In* B. Bolin et al. (ed.) SCOPE 29. The greenhouse effect, climatic change, and ecosystems. John Wiley & Sons, New York.

Shugart, H.H., and W.R. Emanuel. 1985. Carbon dioxide increase: The implications at the ecosystem level. Plant Cell Environ. 8:381–386.

Sionit, N., and P.J. Kramer. 1986. Woody plant reactions to CO_2 enrichment. p. 69–85. *In* H.Z. Enoch and B.A. Kimball (ed.) Carbon dioxide enrichment of greenhouse crops. Vol. 2. Physiology, yield, and economics. CRC Press, Boca Raton, FL.

Sionit, N., D.A. Mortensen, B.R. Strain, and H. Hellmers. 1981a. Growth response of wheat to CO_2 enrichment and different levels of mineral nutrition. Agron. J. 73:1023–1027.

Sionit, N., and D.T. Patterson. 1984. Responses of C_4 grasses to atmospheric CO_2 enrichment. I. Effect of irradiance. Oecologia 65:30–34.

Sionit, N., B.R. Strain, and H.A. Beckford. 1981b. Environmental controls on the growth and yield of okra. I. Effects of temperature and of CO_2 enrichment at cool temperature. Crop Sci. 21:885–888.

Sionit, N., B.R. Strain, and E.P. Flint. 1987. Interaction of temperature and CO_2 enrichment on soybean: Growth and dry matter partitioning. Can. J. Plant Sci. 67:59–67.

Sionit, N., B.R. Strain, H. Hellmers, G.H. Riechers, and C.H. Jaeger. 1985. Long-term atmospheric CO_2 enrichment affects the growth and development of *Liquidambar styraciflua* and *Pinus taeda* seedlings. Can. J. For. Res. 15:468–471.

Smith, S.D., B.R. Strain, and T.D.Sharkey. 1987. Effects of CO_2 enrichment on four Great Basin grasses. Funct. Ecol. 1:139–143.

Solomon, A.M., M.L. Tharp, D.C. West, G.E. Taylor, J.W. Webb, and J.L. Trimble. 1984. Response of unmanaged forests to CO_2-induced climate change: Available information, initial tests, and data requirements. DOE/NBB-0053. U.S. Dep. of Energy, Washington, DC. Available from NTIS, Springfield, VA.

Solomon, A.M., and D.C. West. 1985. Potential responses of forests to CO_2-induced climate change. p. 145–169. *In* M.R. White (ed.) Characterization of information requirements for studies of CO_2 effects: Water resources, agriculture, fisheries, forests and human health. DOE/ER-0236. U.S. Dep. of Energy, Washington, DC. Available from NTIS, Springfield, VA.

Spencer, W., and G. Bowes. 1986. Photosynthesis and growth of water hyacinth under CO_2 enrichment. Plant Physiol. 82:528–533.

St. Omer, L., and S.M. Horvath. 1983. Elevated carbon dioxide concentrations and whole plant senescence. Ecology 64:1311–1314.

Strain, B.R. 1985. Physiological and ecological controls on carbon sequestering in terrestrial ecosystems. Biogeochemistry 1:219–232.

Strain, B.R. 1987. Direct effects of increasing atmospheric CO_2 on plants and ecosystems. Trend Ecol. Evol. 2:18–21.

Strain, B.R., and T.V. Armentano. 1982. Response of "unmanaged" ecosystems. Vol. 2. Part 12. Environmental and societal consequences of a possible CO_2-induced climate change. DOE/EV/10019-12. U.S. Dep. of Energy, Washington, DC. Available from NTIS, Springfield, VA.

Strain, B.R., and F.A. Bazzaz. 1983. Terrestrial plant communities. p. 177–222. *In* E.R. Lemon (ed.) CO_2 and plants: The response of plants to rising levels of atmospheric carbon dioxide. Westview Press, Boulder, CO.

Tissue, D.T., and W.C. Oechel. 1987. Response of *Eriophorum vaginatum* to elevated CO_2 and temperature in the Alaskan tussock tundra. Ecology 68:401–410.

Tolley, L.C., and B.R. Strain. 1984. Effects of CO_2 enrichment on growth of *Liquidambar styraciflua* and *Pinus taeda* seedlings under different irradiance levels. Can. J. For. 14:343–350.

Trabalka, J.R., J.A. Edmonds, J. Reilly, R.H. Gardner, and L.D. Voorhees. 1985. Human alterations of the global carbon cycle and the projected future. p. 247–302. *In* J.R. Trabalka (ed.) Atmospheric carbon dioxide and the global carbon cycle. DOE/ER-0239. U.S. Dep. of Energy, Washington, DC. Available from NTIS, Springfield, VA.

Vengris, J., W.G. Colby, and M. Drake. 1955. Plant nutrient competition between weeds and corn. Agron. J. 47:213–216.

Waggoner, P.E. 1983. Agriculture and a climate changed by more carbon dioxide. p. 383–418. *In* Changing climate: Report of the carbon dioxide assessment committee. National Academy Press, Washington, DC.

Warrick, R.A., R.M. Gifford, and M.L. Parry. 1986a. CO_2, climatic change and agriculture: Assessing the response of food crops to the direct effects of increased CO_2 and climatic change. p. 393–473. *In* B. Bolin et al. (ed.) SCOPE 29. The greenhouse effect, climatic change, and ecosystems. John Wiley & Sons, New York.

Warrick, R.A., H.H. Shugart, M.J. Antonovsky, J.R. Tarrant, and C.J. Tucker. 1986b. The effects of increased CO_2 and climatic change on terrestrial ecosystems. p. 363–392. *In* B. Bolin et al. (ed.) SCOPE 29. The greenhouse effect, climatic change, and ecosystems. John Wiley & Sons, New York.

Washington, W.M., and G.A. Meehl. 1984. Seasonal cycle experiment on the climate sensitivity due to a doubling of CO_2 with an atmospheric general circulation model coupled to a simple mixed-layer ocean model. J. Geophys. Res. 89:9475–9503.

Wigley, T.M.E. 1985. Impact of extreme events. Nature (London) 316:106–107.

Wilks, D.S. 1988. Estimating the consequences of CO_2-induced climatic change on North American grain agriculture using general circulation model information. Clim. Change 13:19–42.

Williams, W.E., K. Garbutt, and F.A. Bazzaz. 1988. The response of plants to elevated CO_2—V. Performance of an assemblage of serpentine grassland herbs. Environ. Exp. Bot. 28:123–130.

Williams, W.E., K. Garbutt, F.A. Bazzaz, and P.M. Vitousek. 1986. The response of plants to elevated CO_2. IV. Two deciduous-forest tree communities. Oecologia 69:454–459.

Wittwer, S.M. 1985. Carbon dioxide levels in the biosphere: Effects on plant productivity. Crit. Rev. Plant Sci. 2(3):171–198.

Wray, S.M., and B.R. Strain. 1986. Response of two old field perennials to interactions of CO_2 enrichment and drought stress. Am. J. Bot. 73:1486–1491.

Wray, S.M., and B.R. Strain. 1987a. Competition in old-field perennials under CO_2 enrichment. Ecology 68:1116–1120.

Wray, S.M., and B.R. Strain. 1987b. Interaction of age and competition under CO_2 enrichment. Funct. Ecol. 1:145–149.

Zangerl, A.R., and F.A. Bazzaz. 1984. The response of plants to elevated CO_2. II. Competitive interactions among annual plants under varying light and nutrients. Oecologia 62:412–417.

8 Implications of Increasing Carbon Dioxide and Climate Change for Agricultural Productivity and Water Resources

J. Goudriaan

Agricultural University
Wageningen, the Netherlands

M. H. Unsworth

University of Nottingham
Loughborough, United Kingdom

ABSTRACT

Green plants need CO_2 to grow. A higher concentration of atmospheric CO_2 will stimulate the photosynthetic process, promoting plant growth and agricultural productivity without increasing the water demand for crop transpiration. On the other hand, expected climatic warming may have adverse effects on agriculture, partly offsetting the positive direct CO_2 effects. The availability of water resources depends on precipitation and potential evaporation, but also on many other factors. Interannual variability cannot be used to extract information of the impact of gradual climatic change. About 5 to 10% of the actual rate of increase of agricultural productivity worldwide can be ascribed to the fertilizing effect of rising atmospheric CO_2. The positive direct effect of CO_2 on plant growth is often smaller when crops are poorly fertilized, but it is fully retained when water shortage limits productivity. Beneficial and detrimental effects of climatic change will not be evenly distributed over the world. Cool and temperate climatic zones will benefit, but in the tropics a further increase in temperature will be undesirable. These changes will exert their influence at such a slow rate that they will be hardly noticeable compared with changes in technology and in economy. Yet they will gradually affect the range of options available.

CLIMATIC EFFECTS

Several atmospheric trace gases such as CO_2, CH_4, and N_2O influence the Earth's climate by transmitting incoming solar radiation, while partly blocking outgoing terrestrial black-body radiation. As the concentrations of

---

 Impact of Carbon Dioxide, Trace Gases, and Climate Change on Global Agriculture, ASA Special Publication no. 53.

these so-called greenhouse gases rise, less terrestrial radiant energy can escape through the Earth's atmosphere. Global surface temperature will rise and restore the radiation balance of the Earth, but a series of interrelated processes that may enhance or reduce this primary effect make it very difficult to predict the final outcome. For example, one important positive feedback process acts by increasing the concentration of water vapor, which itself has a greenhouse effect. On the other hand, a negative feedback may occur through stronger cloud formation (Ramanathan et al., 1989).

The best available estimates now, based on general circulation models, GCM's, predict a global average of 2.5 to 5.5 °K temperature increase, accompanied by some 5 to 15% increase of the circulation rate of the hydrological cycle (Taylor & MacCracken, Chapter 1, this book; Wilson & Mitchell, 1987). We shall briefly review the direct effects of CO_2 and climatic variables on plants and then consider the implications for global agricultural productivity and water resources. Parry et al. (1988) reviewed the subject more broadly.

DIRECT PHYSIOLOGICAL EFFECTS

The trace gas CO_2 is the primary source of C for plant growth. Its present natural concentration is often suboptimal, and for this reason CO_2 enrichment is widely employed in glasshouses to raise yields of crops such as tomato (*Lycopersicon esculentum* Mill.), lettuce (*Lactuca sativa* L.), cucumber (*Cucumis sativus* L.), and flowers (Wittwer & Robb, 1964; Enoch & Kimball, 1986; Nederhoff & van Uffelen, 1988).

The major physiological effect of increased CO_2 is to increase the rate of CO_2 assimilation. Generally, the utilization efficiency of other plant growth factors (e.g., radiation, water, nutrients) is also improved. These direct physiological effects are considered by Acock (Chapter 4, this book) and by Idso (Chapter 5, this book). A large body of research in this field was brought together in the thorough review edited by Strain and Cure (1985) for the U.S. Department of Energy. Likewise, reports have been prepared in the Netherlands (Gezondheidsraad, 1983, 1987) and in Great Britain (Squire & Unsworth, 1988). For this reason we can permit ourselves to be brief, but for clarity some repetition cannot be avoided.

Interactions and partial compensations of direct and indirect effects may occur. The results of some crop growth models that allow for these compensations will be discussed.

DIRECT CARBON DIOXIDE EFFECTS ON LEAF, CANOPY, AND CROP

Leaf Response of Photosynthesis

A large body of evidence is available on experimentally established effects of ambient CO_2 (C_a) on the photosynthesis of leaves and crop canopies, and on the subsequent dry matter accumulation in several crop components (Lemon, 1983; Strain & Cure, 1985; Kimball, 1983; Cure & Acock, 1986; Acock, Chapter 4, this book). Light is indispensable for the process of photosynthetically driven CO_2 uptake by green plants. In the natural environment, light and CO_2 are normally suboptimally present; consequently, photosynthesis is stimulated by an increase of ambient CO_2 (Fig. 8-1), not only under high light, but also under low light conditions.

There are two major classes of plants called C_3 and C_4 plants, which differ in biochemical and anatomical aspects of the way they take up CO_2 from the ambient air (see Acock, Chapter 4, this book). In C_3 plants the enzyme that binds CO_2 with RuBP, can also bind RuBP with O_2, after which RuBP must be recovered (Farquhar & von Caemmerer, 1983). This recovery costs energy and releases CO_2, observable as photorespiration. Because CO_2 and O_2 compete for the same site on the catalyzing enzyme, photorespiration is suppressed by higher CO_2. In the other direction, when CO_2 is lowered to the level where CO_2 uptake and CO_2 release just balance each other, this concentration is called the CO_2 compensation point, Γ (Fig. 8-2). The value of Γ is about 50 μmol mol^{-1} at 25 °C and rises with increasing temperature. The level of Γ is proportional with O_2 concentration. The light-saturated rate of photosynthesis is proportional to ($[CO_2] - \Gamma$) up to

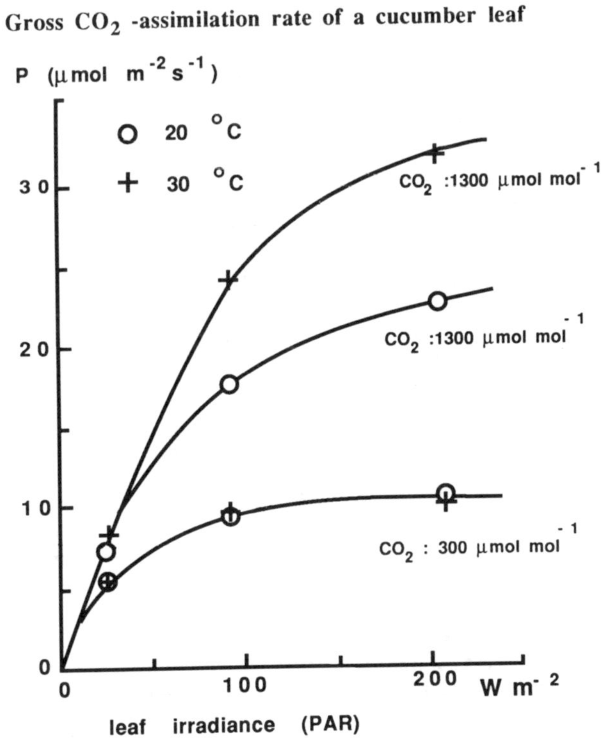

Fig. 8-1. Typical photosynthesis-light response curves, as stimulated by higher CO_2 (after Gaastra, 1962).

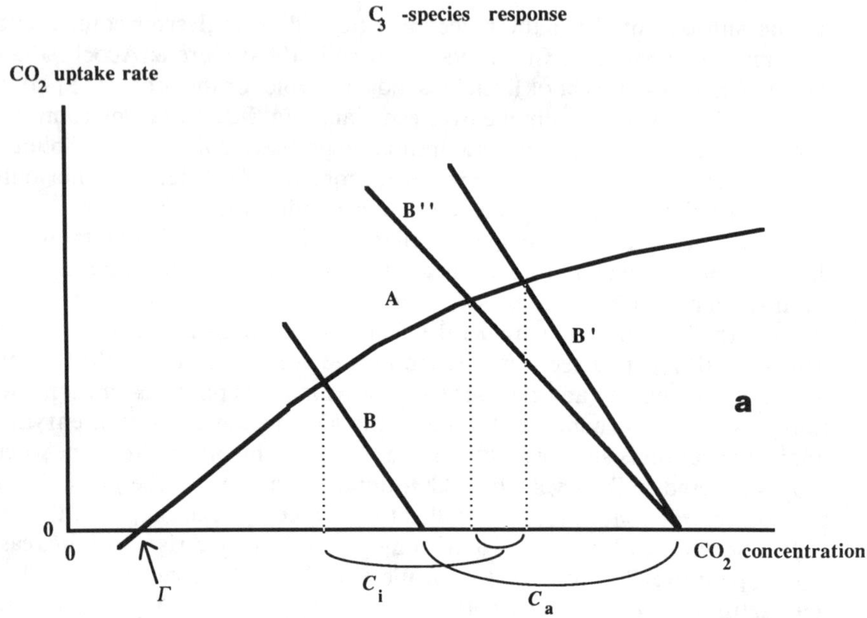

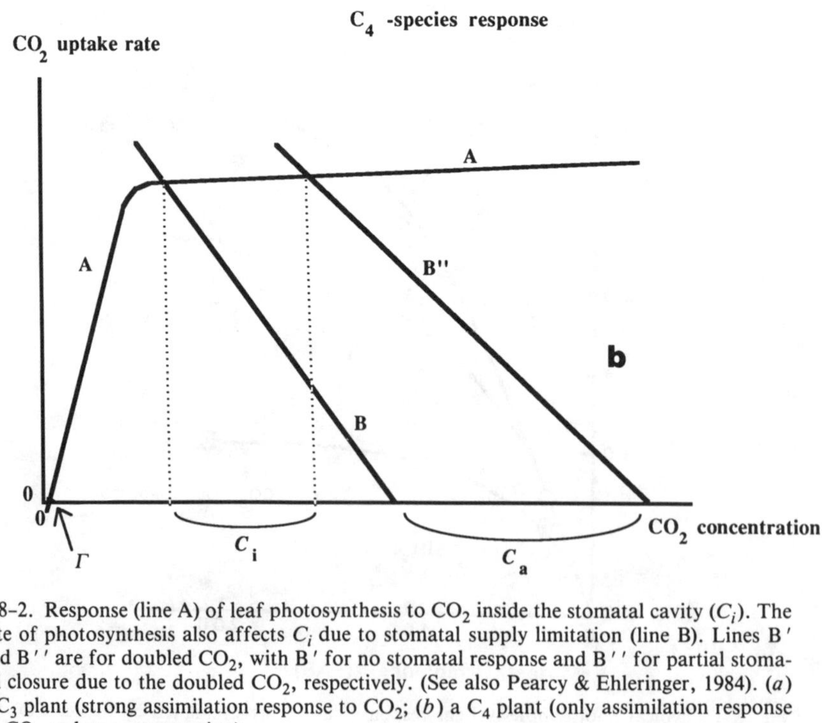

Fig. 8–2. Response (line A) of leaf photosynthesis to CO_2 inside the stomatal cavity (C_i). The rate of photosynthesis also affects C_i due to stomatal supply limitation (line B). Lines B′ and B′′ are for doubled CO_2, with B′ for no stomatal response and B′′ for partial stomatal closure due to the doubled CO_2, respectively. (See also Pearcy & Ehleringer, 1984). (*a*) a C_3 plant (strong assimilation response to CO_2; (*b*) a C_4 plant (only assimilation response to CO_2 at low concentration).

a concentration of about 1000 μmol mol^{-1}, and the initial light use efficiency increases with CO_2 in proportion with $([CO_2] - \Gamma)/([CO_2] + 2\Gamma)$ (Goudriaan et al., 1985).

In C_4 plants {mostly tall tropical grasses such as millet (*Panicum miliaceum* L.), maize (*Zea mays* L.), sorghum [*Sorghum bicolor* (L.) Moench] and sugarcane (*Saccharum officinarum* L.)} not RuBP but PEP, which does not bind with O_2, is used as a preliminary binder of CO_2 from the external air. The bound CO_2 is transported from the mesophyll cells to the sheath bundle cells in the form of organic acids and then released. The RuBP-CO_2 reaction is concentrated in these bundle sheath cells, where a very high CO_2 concentration is maintained. Oxygen gets practically no chance to bind with RuBP, so that photorespiration is suppressed and Γ stays low (about 5 μmol mol^{-1}). Because of the high affinity of PEP to CO_2, the photosynthetic response to CO_2 is very steep until about 100 μmol mol^{-1}. In the normal range of CO_2 of 300 to 500 μmol mol^{-1}, the C_4 leaf assimilation rate rises only little with CO_2, even under high light intensities.

Leaf Response of Stomatal Conductance

A typical secondary effect of raising C_a is partial stomatal closure, which increases the exchange resistance between leaf and ambient air to gases such as CO_2 itself and water vapor (Raschke, 1975; Morison, 1987). The presence of stomatal pores in the epidermis of leaves permits uptake of CO_2 from the ambient air, but at the same time water vapor escapes, leading to transpiration. The degree of opening can be considered as a compromise in the balance between limitation of water loss and admission of CO_2. The generally observed strong correlation between stomatal conductance and demand for CO_2 is an expression of this compromise (Wong et al., 1979).

Not only is stomatal conductance well correlated with photosynthesis when light varies (Goudriaan & van Laar, 1978), but also when unfavorable conditions occur such as air pollution, shortages of water or nutrients (Goudriaan & van Keulen, 1979), or senescence (Goudriaan & van Laar, 1978). The much higher affinity for CO_2 in C_4 plants permits them to maintain a more favorable ratio between net CO_2 uptake and stomatal conductance. Typically the concentration ratio of intercellular CO_2 (C_i) to ambient CO_2 is about one-third in C_4 plants, but about two-thirds in C_3 plants.

When ambient CO_2 is raised, net CO_2 assimilation may be increased and/or water loss may be reduced, depending on how the stomata react. In Fig. 8–2 the supply-demand function of assimilation and CO_2 is given. The figure shows how the assimilation rate responds to intercellular CO_2 concentration (line A). The intercellular concentration declines from C_a linearly when assimilation rate increases (line B), due to the resistance offered by the stomata and leaf boundary layer. The intersect of the two lines (line B with line A) gives the realized rate of assimilation and of intercellular CO_2 concentration. Raising external CO_2 will cause a maximal assimilation effect (line B ′) if stomata do not respond at all, but usually the stomata close somewhat (line B ′ ′). In the extreme situation of constant intercellular CO_2

the stomatal effect will be maximal, but the assimilation effect will have vanished. The most common response is that the ratio of intercellular to external CO_2 concentration is stabilized, thereby partitioning the effect between assimilation and water loss. This constant C_i/C_a ratio means that stomatal aperture is reduced very little if assimilation responds strongly to C_i (typical in C_3 plants). In C_4 plants photosynthesis uptake is usually almost CO_2 saturated, and stomatal closure is maximal (Fig. 8–2b).

A compilation of data by Morison (1987) showed that stomatal conductance at $C_a \approx 660$ µmol mol^{-1} was reduced to about 60% on average compared with the value at 330 µmol mol^{-1}. The variability around this reduction resulting from varying responses of assimilation to doubled CO_2, ranging from none in C_4 plants to about 40% in C_3 plants, is not large. In both groups water use efficiency (WUE) is considerably stimulated by increased ambient CO_2, either by stimulated assimilation rate or by reduced transpiration rate. Typically in C_3 plants transpiration will be reduced by 10 to 20% and assimilation stimulated by 40%, and in C_4 plants only transpiration will be reduced, up to about 25%.

Canopy Response

Leaf responses as described above are also found in well-ventilated canopies (Fig. 8–3, after Louwerse, 1980).

Doubt has been raised (Jarvis & McNaughton, 1986) about whether a CO_2-induced reduction of transpiration can be maintained on the scale of

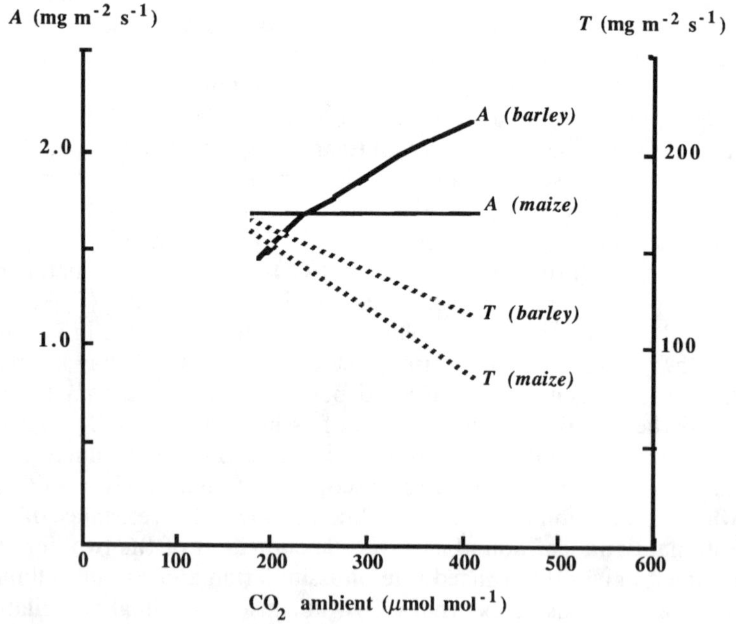

Fig. 8–3. Measured canopy responses to ambient CO_2 of photosynthesis (A) and transpiration (T) in barley (*Hordeum vulgare* L.) (C_3) and maize (C_4) (after Louwerse, 1980).

extended vegetated surfaces, because negative feedback by increased leaf-air humidity gradients may nullify the initial gain. Such a negative feedback will be particularly strong when there is a poor coupling with the atmosphere in conditions of low wind and/or an aerodynamically smooth crop surface.

This is, however, only part of the story. Precisely in those circumstances of large boundary layer resistance, the canopy air tends to be depleted in CO_2. Increased atmospheric CO_2 will then be effective in removing this depletion limitation and in stimulating assimilation rate. As a result, although transpiration will be reduced less, assimilation will be stimulated more than initially expected.

Microweather simulation enables an investigation of these effects on a canopy scale. Rosenberg et al. (1990) calculated that a 40% increase in stomatal resistance reduced evapotranspiration between 9% [in wheat (*Triticum aestivum* L.)] and 18% (in grassland), which shows that the effect of stomatal resistance is considerably reduced but not nullified.

In this chapter, results are presented from a stratified micrometeorological model (Goudriaan, 1977; Chen, 1984), which also included CO_2 effects on photosynthesis (Goudriaan et al., 1985). This model includes detailed in-canopy profiles of radiation and of aerial conditions, using first-order turbulent transport theory. The leaf energy balance and soil surface energy balance are based on the Penman-Monteith combination equation. Exchange above the canopy is modelled according to similarity for wind, mass, and heat, corrected for thermal gradients.

The results (Table 8-1) showed that in the more realistic simulation (Case b), the reduction of transpiration by doubled CO_2 was less than in the over-simplified run (Case a), where the micrometeorological feedback was ignored. Interestingly, the micrometeorological feedback enhanced the positive CO_2 effect on assimilation. This simulation included almost all steps in the scaling-

Table 8-1. Simulated effects of doubled CO_2 on daily assimilation (A), on daily transpiration (T), and on water use efficiency (WUE) for C_4 and C_3 plants, calculated by the Goudriaan-Chen model. In the control situation, crop height was 0.2 m and leaf area index was fixed at 4.

	C_4			C_3		
$[CO_2]$	A	T	WUE	A	T	WUE
μmol mol^{-1}	g m^{-2} d^{-1}	kg m^{-2} d^{-1}	g kg^{-1}	g m^{-2} d^{-1}	kg m^{-2} d^{-1}	g kg^{-1}
A. With air conditions defined at a height of 0.2 m (no micrometeorological feedback)						
350	88.03	3.516	25.04	63.988	4.562	14.03
700	88.03	2.265	38.87	96.477	3.729	25.87
Ratio	1	0.644	1.55	1.508	0.817	1.84
B. With air conditions defined at a height of 20 m (so that there was micrometeorological feedback)						
350	88.03	3.631	24.24	61.03	4.074	14.98
770	88.03	2.699	32.62	94.39	3.654	25.83
Ratio	1	0.743	1.35	1.547	0.897	1.72

up process mentioned by Jarvis and McNaughton (1986). Only the step from large field to region was not included.

In C_4 species, the simulated assimilation rate was not affected, but the stomatal resistance was almost proportional to external CO_2. Indeed, the simulated crop resistance around 1200 h (noon) increased from 70 to 127 s m^{-1}, which is a reduction to 55% in terms of crop conductance. The transpiration rate, however, did not decrease in the same proportion: it was reduced to 64% when microclimatic air conditions inside the canopy were kept the same. This stabilization of transpiration rate was caused by increased leaf temperature. When not only leaf temperature, but also the microclimatic air conditions inside the canopy were affected, transpiration crept up further to reach 74% of the original value.

For C_3 species the major CO_2 effect was in assimilation rather than in transpiration. In spite of the micrometeorological feedback in all simulated circumstances, however, a considerable improvement of water use efficiency was found, varying between 1.3 for C_4 species to 1.7 for C_3 species. These results suggest that improvement of WUE is a more robust effect of CO_2 than its antitranspirant effect (Allen et al., 1985).

Crop Responses

During the growing season, several internal mechanisms operate in the plant modifying the initially observed effects of CO_2. In plants adapted to high CO_2, photosynthesis per unit leaf area is often smaller (Wong, 1979; Mortensen, 1983) than that of nonadapted plants (when measured under equal circumstances), but in soybean [*Glycine max* (L.) Merr.] stimulation of photosynthetic capacity has been observed (Valle et al., 1985; Campbell et al., 1988). When both grown and measured at a higher CO_2 concentration, leaves have generally a higher rate of photosynthesis. This is particularly true for N_2 fixing plants, which have nodules in their rooting system. The growth response of these plants to CO_2 tends to be particularly strong.

Starch accumulation (Ehret & Jolliffe, 1985) tends to cause some increase of leaf weight per unit leaf area. This rather passive response will not increase light interception, but the more active response of more leaf area formation by larger leaves or more tillers will further enhance the effect of increased CO_2 when considered at a crop level.

As shown in a review by Kimball (1983), responses at the single leaf level to CO_2 are carried over to crop yield and can be summarized by a mean 40% increase of dry matter for C_3 crops upon doubling of CO_2, and by 15% for C_4 crops.

A similar overall response for a C_3 crop was simulated by a fairly complicated model (Goudriaan et al., 1985), which considers the physiological mechanisms at the leaf level. More specifically, the model showed that the overall C response could be well-described by the following logarithmic equation (Fig. 8–4), at least in the domain of about 200 to 1000 μmol mol^{-1}:

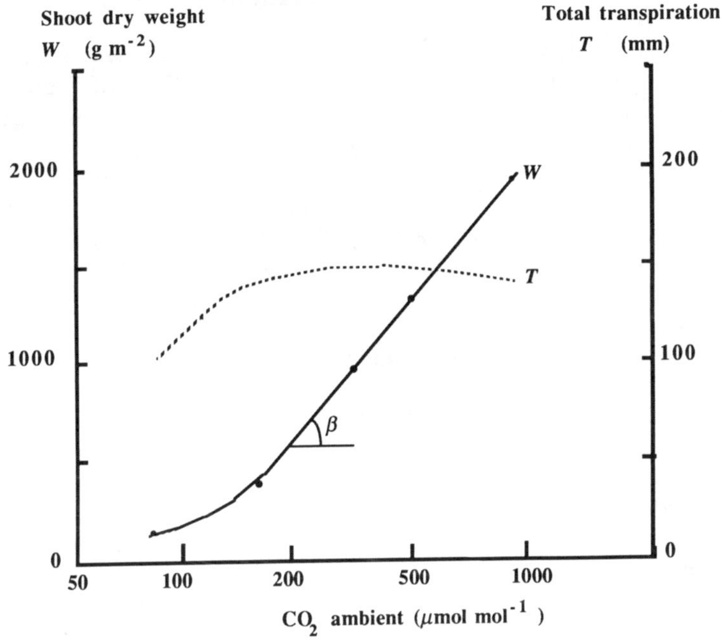

Fig. 8–4. Simulated total dry matter and whole season water consumption as affected by ambient CO_2, for a crop of Faba bean (Goudriaan et al., 1985).

$$\frac{W}{W_0} = 1 + \beta \ln \left(\frac{C}{C_0}\right) \qquad [1]$$

In this equation the subscript 0 stands for the reference situation, W for crop dry weight at the end of the season, and C for ambient CO_2 concentration. This equation is only indicative and has no physiological meaning, but it serves well to summarize many observations and simulation results. The response factor β of modelled crop dry weight was found to be 0.7 (see also Allen et al., 1987) when the increased C gain was invested again and used for increased tillering, increased leaf number, or increased leaf size. In this simulation run, total transpiration over the growing period did not decrease upon doubling of CO_2, in spite of stomatal closure, because the modeled increase of leaf area compensated for decreased leaf transpiration. Several experimental investigations have reported this type of stability of total crop transpiration (Jones et al. 1984; Kimball et al., 1983, 1984).

When no positive feedback on leaf area growth was assumed, the additional C was stored as starch in leaf or tuber, and the value of β was smaller, about 0.5. In this case the modeled cumulative transpiration did decrease in doubled CO_2.

INTERACTION WITH WATER SHORTAGE

The positive effect of CO_2, observed under good growing conditions, is fully maintained under water shortage. Gifford (1979) and Sionit et al. (1980) have shown that water relations, growth, and yield of wheat under water shortage can be considerably improved by raised ambient CO_2. Kimball et al. (1986) reported relative increases of cotton (*Gossypium hirsutum* L.) production ranging between 50 and 70% under doubled CO_2, in both optimal and limiting levels of water supply. Goudriaan and Bijlsma (1987) showed that WUE in Faba bean (*Vicia faba* L.) was improved by about 50% under doubled CO_2, for both normal and limited water supply. For a C_4 grass, however, Gifford and Morison (1985) found that growth stimulation by CO_2 began only under severe water stress.

In saline situations plants grow continually under osmotic stress. Schwarz and Gale (1984) found that the CO_2 effect on dry matter accumulation was equally strong or even more promounced in saline than in nonsaline situations, and that C_4 halophytes (plants that are salt-tolerant) responded as equally strongly as C_3 halophytes.

A major reason for the strong positive CO_2 effect under water stress, clearly, is the common physical pathway of water vapor and CO_2 through the stomatal pores. Anothcı partial explanation can be found in better availability of assimilates for making osmotic adjustments and thereby maintaining turgor. These results are especially important for arid regions, where brackish or even saline situations are common.

When a limited stock of water is available, the plants may have adequate water most of the growing period. The water shortage only occurs by the time the soil has lost about 75% of its initial water content. The increased WUE as mentioned earlier will result in proportionally more biomass, but the amount of water left in the soil will be the same. However, if colimitation by nutrients or by crop development duration occurs, the increase in biomass will be too small to compensate for the decreased transpiration rate and the amount of soil water left after the growing period may increase.

INTERACTION WITH NUTRIENT SHORTAGE

Nutrient shortage tends to impose more strongly a limitation to crop growth than water shortage, without leaving much room for stimulation by CO_2. Increased starch accumulation in leaves grown under high CO_2 is responsible for a general increase in dry matter.

Phosphorus shortage was found to limit growth almost independently of CO_2 (350 and 700 μmol mol^{-1}) in a pot experiment (Goudriaan & de Ruiter, 1983). Also, the growth-limiting effect of K shortage was not alleviated by higher CO_2 (J. Goudriaan, 1985, unpublished data).

In open soil and perennial species a positive interaction may develop, by more intense rooting and increased soil weathering (Rosenberg, 1981). Norby et al. (1986) could not find a disappearance of the positive CO_2 effect, even at very low nutrient levels.

Nitrogen appears to be in an intermediate position, in allowing a small positive CO_2 effect, even under rather severe N shortage (Goudriaan & de Ruiter, 1983). In addition to starch accumulation, which generally lowers leaf nutrient contents, there is also a decrease in the carboxylation enzyme Rubisco, which contains up to 50% of leaf N. In accordance with this photosynthetic role of N, the N content in leaf tissue is lower for plants grown under high CO_2 than under low CO_2. In seed tissue, however, the C/N ratio is more stable (Kimball et al., 1986).

An intriguing question is how canopy transpiration is affected by CO_2 when plant growth is nutrient-limited. Lenssen (1986) showed that the transpiration reduction by higher CO_2 was still fully present under severe P limitation. This result is relevant for runoff from watersheds with naturally growing vegetation. Quite often these natural ecosystems are strongly nutrient-limited. When growth is not improved under increased CO_2, some increase of the water yields may be expected.

INTERACTION WITH CLIMATIC EFFECTS

Temperature and the Growing Season

In the absence of other limiting factors, temperature in cooler climate regions determines the length of the growing season, principally by its effects on the timing of developmental processes and on rates of expansion of leaves. Most processes of development and expansion (e.g., leaf expansion, flowering, grain filling) can be described by a linear increase of rate with temperature from a minimum threshold temperature to an optimum, and then a linear decrease to a maximum temperature limit, above which the process ceases (Monteith, 1981).

The implications of the temperature response for crop production depend on whether the growth form is *determinate* (as in cereals that have a discrete life cycle, ending when the crop is mature) or *indeterminate* (as in perennial grass or in potato, *Solanum tuberosum* L., which continues growing and forming yield as long as the temperature is above the minimum threshold). In determinate crops of temperate regions, higher temperatures decrease the time for which the canopy exists, and hence, decrease radiation interception and decrease potential yield. In contrast, higher temperatures *increase* the potential yield of indeterminate crops.

When higher temperatures are increased along with increased CO_2 concentrations, as in greenhouse effect scenarios, the two factors act in opposite directions for C_3 cereals. Squire and Unsworth (1988) used a mechanistic model of winter wheat growth to assess the impact of CO_2 and temperature on potential wheat yields in England. They assumed that water, nutrition, pests, and disease were not limiting factors.

When CO_2 concentration was doubled, but daily weather data from a typical year (1981) were used, the potential grain yield was 27% larger than in a control run at 340 μmol mol^{-1} CO_2 (11.5 and 9.0 Mg ha^{-1}, respectively). The date of maturity was unchanged. When daily temperatures were increased by 3 °C *and* CO_2 was doubled, more rapid development of the crop shortened the growing season, and the potential grain yield (10.4 Mg ha^{-1}) was only 15% larger than in the control, but maturity occurred 30 d earlier. A combination of +4.5 °C and doubled CO_2 resulted in the temperature and CO_2 factors almost cancelling out, so that the potential grain yield (9.7 Mg ha^{-1}) was only 8% larger than the control. Maturity was 40 d earlier than in 1981 conditions.

On the other hand, in some conditions also in determinate species a positive interaction with temperature exists. Maize in northwestern Europe does not reach maturation, but it is used as a silage crop instead. Because its fodder value is largely determined by the fraction of cobs, even when not mature, a longer growing season is highly profitable. Moreover, in these cool regions maize usually grows below its optimal temperature for photosynthesis.

In most regions in North America (Rosenzweig, 1985), and also in southern Europe, increased temperature would lead to losses of crop yield; the same conclusion applies to the tropical and subtropical regions of the world. Moreover, especially in the tropics, agriculture tends to suffer from nutrient shortage, which further impairs the potential to utilize the beneficial direct effects of increased CO_2.

Negative compensatory effects due to water shortage in midsummer may still occur. Also, pests and diseases may become more serious, because more time will be available for them to generate larger populations during the growing season, and because of greater survival through the winter.

Other Climatic Factors

Other climatic factors such as precipitation and radiation are no less important than temperature (Rosenberg et al., 1990). Unfortunately, there is even less certainty about the way they will change than there is for temperature. Crop growth models show that the potential productivity is closely related to incoming radiation during the growing season, but the GCM's are not able to produce reliable predictions on this point yet. Therefore, the possible impact of radiation change is usually ignored, in spite of its potential importance.

For precipitation, the currently available figure is that the mean global precipitation will increase by 7 to 15% (Wilson & Mitchell, 1987). Some regional specification can be obtained from maps such as produced by Schlesinger and Mitchell (1985). A model for wheat and rice (*Oryza sativa* L.) that included both the climatic (temperature and precipitation) effects and the physiological effects of doubled CO_2 (van Diepen et al., 1987), predicted potential crop yields to increase by 10 to 50% for some sites in Europe and Asia. The mean GCM scenarios were superimposed on current weather, retaining current variability. The resulting yield variability was not increased, except in the Mediterranean zones.

In another simulation study for crop yields in the USA, Curry et al. (1988) used climate scenarios based on GCM studies with the GISS model and GFDL models (Taylor & MacCracken, Chapter 1, this book). However, the precipitation scenarios from these models were quite different, and also showed, month by month, extremely uneven precipitation changes. Further climate study will be needed to confirm if such extreme changes are realistic.

SOME HYDROLOGICAL CONSIDERATIONS

A Simple Equation for Runoff

River runoff is one of the major factors for the availability of water resources (da Cunha, 1988). Runoff is precipitation that has not been either evaporated or stored in the soil. Collected over an entire river catchment area, it gives rise to the river discharge. Because river discharge is the result of spatially heterogeneous and time-varying contributions from subcatchment areas, complicated models are required for its complete description (Gleick, 1986, 1987).

Here a rudimentary analysis will be given based on a few simple equations. Averaged over the annual cycle, soil storage can be neglected and the runoff, R, is simply the difference between precipitation, P, and actual evapotranspiration, E:

$$R = P - E \qquad [2a]$$

The type of vegetation can have a considerable effect on E, even under the same climatic conditions (Dickinson & Henderson-Sellers, 1988). For example, Dolman and Nonhebel (1988) showed that a spruce (*Picea* sp.) forest loses much more water than an oak (*Guercus* sp.) forest due to direct evaporation of intercepted water. In fact, for long-term studies the change of vegetation type along with climate should also be taken into account. This problem can be partly solved by using an equation that includes covariation of vegetation with climate. For that purpose we use the following simple, old empirical equation for actual annual evapotranspiration, E, based on annual precipitation, P, and potential evapotranspiration E_p. This equation goes back to Oldekop (1911) (Dooge, 1988):

$$E = E_p \tanh (P/E_p) \qquad [2b]$$

This equation does not include the water holding capacity of the soil nor the effect of vegetation on any other surface parameterizations. It does, however, follow the basic physical requirements that E approaches to P for very low P/E_p on one hand, and to E_p for very high P/E_p on the other. The equation also performs well when P and E_p are of the same order (Table 8–2).

As shown in Table 8–2, mean runoff is well predicted. In the polders in the Netherlands (de Vries, 1980), not only precipitation (750 mm yr^{-1}), but also seepage water and inlet water (180 mm yr^{-1}), contribute to the

Table 8-2. Comparison of the equation $R = P - E_p \tanh (P/E_p)$ and observed data for a few regions.

Region	P	E_p	$R_{predicted}$	$R_{observed}$
	mm yr^{-1}			
Eastern USA (Schaake, 1990)	1140	890	378	355–410
Upper Colorado (Revelle & Waggoner, 1983)	350	450	57	53
Rhine catchment (Thran & Broekhuizen, 1965)	800	600	278	280
Polders in Holland (de Vries, 1980)	750	600	241	440
With seepage and inlet	930	600	382	440

positive side of the water balance, in total 930 mm yr^{-1}. Potential evaporation is 600 mm yr^{-1}, so that R is predicted to be 382 mm yr^{-1}, close to the 440 mm yr^{-1} that is pumped out in reality.

The behavior of R, based on Eq. [2a] and [2b], is graphed in Fig. 8-5 by isolines for R as a function of P and E_p. Two widely different locations are indicated in the graph, an "R" for the river Rhine catchment area, and a "C" for the Colorado River catchment area. The major river in northwestern Europe, the river Rhine, has a catchment area of about 220 000 km^2, and a mean discharge of 2000 m^3 s^{-1}. According to the maps compiled by Thran and Broekhuizen (1965), the mean precipitation over the catchment area is 800 mm yr^{-1} and the mean potential evaporation is 600 mm yr^{-1}, so that Eq. [2] predicts a mean runoff of 278 mm yr^{-1}. This leads

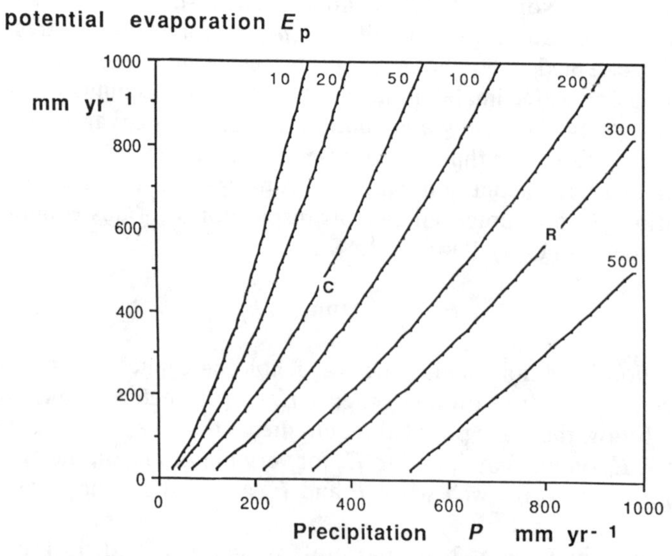

Fig. 8-5. Isolines of equal runoff as a function of precipitation, P, and potential evaporation, E_p, calculated with Eq. [2]; C and R characterize the catchment areas of the Upper Colorado and the Rhine River, respectively.

to a predicted river discharge of 1940 $m^3 s^{-1}$, in excellent agreement with the data. For the Colorado River we used the publication of Revelle and Waggoner (1983), who give a P of 350 mm yr^{-1} and an E_p of 450 mm yr^{-1}, based on Langbein et al. (1949). Equation [2] then gives 57 mm yr^{-1} as discharge, which compares very well with the 53 mm yr^{-1} mentioned by Revelle and Waggoner.

This equation is useful because of its simplicity, enabling a quick assessment of the sensitivities of river discharge for changes in precipitation and potential evaporation. According to the GCM studies, the expected global warming will be accompanied by an increased hydrological cycle. Averaged over the whole globe, P and E_p will increase by 7 to 15% upon doubling of atmospheric CO_2. A robust first-order conclusion is that the runoff ratio R/P will not alter if P and E_p increase by the same relative fraction. The mean runoff R itself will then also increase by about 7 to 15%. Regionally, of course, P and E_p may change in a different way. If just P and not E_p increases, runoff will increase much more than proportionally, and vice versa. According to Eq. [2], the relative sensitivity of R to P, $(\Delta R/R)/(\Delta P/P)$, is as high as 3 in very dry regions ($P/E_p \ll 1$), and drops to about 2.4 at $P/E_p = 1$ and 1.8 at $P/E_p = 2$. If, on the other hand, only E_p increases, R to E_p will be diminished with a relative sensitivity, $(\Delta R/R)/(\Delta E_p/E_p)$, of -2 in very dry regions, -1.4 in moderately humid zones ($P/E_p = 1$), and -0.8 in very humid conditions ($P/E_p = 2$).

In most temperate climates, precipitation and potential evaporation are of similar magnitude. According to Eq. [2] the runoff is then about 25% of the precipitation, and the relative sensitivities of R to P and E_p are 2.4 and -1.4, respectively.

Climate Change and Annual Precipitation/Potential Evaporation Ratio

The simplest explanation for the contrast of summer dryness and winter wetness, at least in temperate and mediterranean climates of the world, is the following. In winter the oceans are still relatively warm and can give moisture to the air above them. This humid air spreads out and the relatively cold continents receive a surplus of winter precipitation, especially at their margins. Because of the low temperatures, almost no moisture is returned to the air, so the center regions of the continents stay cold and dry. In summer, on the other hand, the surface temperature of the sea lags behind that of the land. The precipitation surplus at the continents disappears, but now there is a large evapotranspiration from the land masses. In combination with the large moisture-carrying capacity of the warm air, precipitation is now supplied to the continental center regions as well.

Clearly, dryness and temperature are highly correlated when they are both followed over the seasons. Why, then, should there not be more dryness when temperature rises due to the greenhouse effect? There is no reason to assume that the annual temperature cycle, and in particular the phase lag between sea and land, will become more extreme. The greenhouse warming will be so slow that, to a first approximation, both cycles at land and sea will go up together. Therefore, a generally increased dryness is not to be expected. Temporarily the sea temperature may show some lag, but with a magnitude that is likely to be negligible in comparison to the annual cycle.

This is not to say that locally no changes in climate can occur, but even then it should be kept in mind that the geographical distribution of aridity patterns is largely governed by mountain ranges and by cold and warm ocean currents. These currents of course may shift, which poses the largest danger. For northwestern Europe the Gulf Stream is vital, and it would be a disaster if it disappeared or even changed its course. The importance of sea surface temperature was recently confirmed by Palmer and Brankovic (1989).

Direct Carbon Dioxide Effects and Annual Precipitation/Potential Evaporation Ratio

In extremely wet climates, the evapotranspiration will not be affected by CO_2. Vegetation and soil are wet all the time, and almost all evaporation is direct evaporation of intercepted water, in which stomatal opening plays no role. But also in extremely dry climates the role of vegetation is negligible, and runoff is largely governed by excess water flowing from the bare ground after an occasional storm. A better WUE of vegetation will help the vegetation survive or increase its biomass, but the water yield from the catchment area will not rise. Conceivably, an increase in vegetation could even reduce water yield in these arid regions by reducing the ease with which water runs off after the occasional rains.

In the important intermediate situation where P and E_p are of similar magnitude, some reduction of actual transpiration may occur. The results discussed earlier, however, indicate an increased leaf area will compensate for decreases in stomatal opening. A purely empirical equation such as Eq. [2b] cannot account for such effects, and detailed mechanistic models will be necessary. Our expectation is, however, that such effects will be small, and that changes in type of land use will be much more important.

CONCLUSIONS

Implications for Global Agriculture

A rather detailed simulation experiment by Wilks (1988), using a physiological process-based simulation model for cereal growth, in combination with GCM based information on climate change in a double-CO_2 world indicated that for the USA Corn Belt the expectation for crop yield is not dominantly negative. Both positive and negative effects for wheat and corn (*Zea mays* L.) were predicted, ranging between −40% and +80%. The direct physiological effects of CO_2 on photosynthesis and stomatal opening were not included, however. Similarly, Parry et al. (1988) and Rosenzweig (1985) did not consider the direct effects of CO_2, and it seems safe to consider these results as "worst case" scenarios. As explained, the direct effects of CO_2

for crop production are generally beneficial and can be added to the indirect climatic effects. Direct and indirect effects together will probably show a predominantly positive picture.

Over the period 1959 to 1986, almost 30 yr, the world average grain yield per unit area has increased from about 1400 kg ha^{-1} to about 2600 kg ha^{-1} (FAO, 1987). This near-doubling is the result of a continued annual relative growth rate of about 2.3% per year. At the same time, atmospheric CO_2 has risen from 315 to about 345 μmol mol^{-1}, at an average relative rate of about 0.34% yr^{-1}. When plotted together the correlation is almost perfect (Fig. 8–6). It would be very unwise indeed to draw any conclusions from this graph as far as cause and effect are concerned. Rather, it brings us to some interesting considerations.

It shows that the relative growth rate of cereal production (on an area basis) has been seven times as large as that of atmospheric CO_2. Even under optimal conditions for growth, the socalled β factor of CO_2 will rarely exceed the value of 0.5, which means that at most 1/14 of the actual increases in crop production can be reasonably ascribed to increasing atmospheric CO_2. Is this much? It is indeed, if one considers it as a free premium, on which we can reckon in the foreseeable future. It is not large, however, if it is compared with the human dynamics in the agricultural technology. In fact, its relatively small impact confirms the point of view of those who consider the climatic change issue as minor when compared with other opportunities and dangers of a man-made nature. The slowness of the steady drift

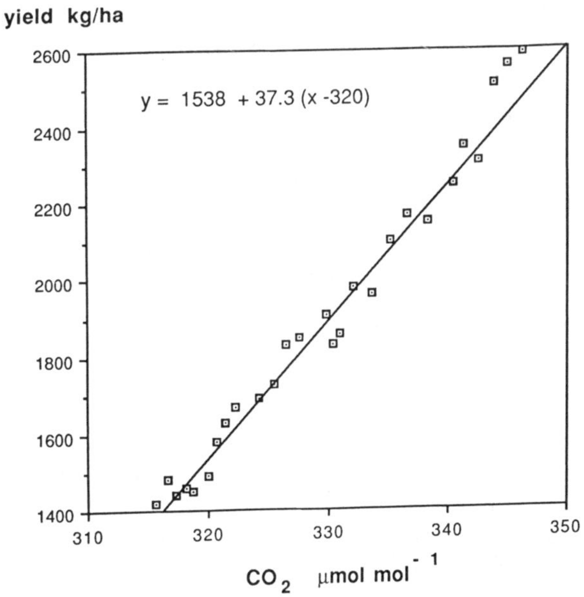

Fig. 8–6. World average grain yields over the period 1959 to 1986 (kindly supplied by H. van Heemst at CABO, Wageningen from FAO production yearbooks) vs. atmospheric CO_2, measured at Mauna Loa, HI.

of the environment permits ample time for adaptation, even for time-demanding activities such as plant breeding (Wittwer, 1980).

There is, however, also a danger in this point of view, namely that it is too general. There may be parts of the world where agriculture may be even completely pushed out of its borders of existence, most literally by sea-level rise, but also by deterioration of already extreme climatic conditions.

REFERENCES

Allen, L.H., K.J. Boote, J.W. Jones, P.H. Jones, R.R. Valle, B. Acock, H.H. Rogers, and R.C. Dahlman. 1987. Response of vegetation to rising carbon dioxide: Photosynthesis, biomass, and seed yield of soybean. Global Biogeochem. Cycles 1:1–14.

Allen, L.H., P. Jones, and J.W. Jones. 1985. Rising atmospheric CO_2 and evapotranspiration. p. 13–27. *In* Advances in evapotranspiration ASAE Publ. 14–85. ASAE, St. Joseph, MI.

Campbell, W.J., L.H. Allen, and G. Bowes. 1988. Effects of CO_2 concentration on Rubisco activity, amount, and photosynthesis in soybean leaves. Plant Physiol. 88:1310–1316.

Chen, J. 1984. Uncoupled multi-layer model for the transfer of sensible and latent heat flux densities from vegetation. Boundary-Layer Meteorol. 28:213–226.

Cure, J.D., and B. Acock. 1986. Crop responses to carbon dioxide doubling: A literature survey. Agric. For. Meterol. 38:127–145.

Curry, R.B., R.M. Peart, J.W. Jones, K.J. Boote, and L.H. Allen, Jr. 1988. Simulation as a tool for analyzing crop response to climate change. ASAE Pap. 88-7512. ASAE, St. Joseph, MI.

da Cunha, L.V. 1988. Climatic changes and water resources development. p. 639–660. *In* A. Berger et al. (ed.) Climate and geo-sciences. NATO ASI Ser. C. Vol 285. Kluwer Academic Publ., Dordrecht, Netherlands.

de Vries, J.J. 1980. Hydrologie van Nedereland. Rodopi, Amsterdam, Netherlands. (in Dutch).

Dickinson, R.E., and A. Henderson-Sellers. 1988. Modelling tropical deforestation: A study of GCM land-surface parameterizations. Q. J. R. Meteorol. Soc. 114:439–462.

Dolman, A.J., and S. Nonhebel. 1988. Modelling forest water consumption in The Netherlands. Agric. Water Manage. 14:413–422.

Dooge, J.C.I. 1988. The role of the hydrological cycle in climate. p. 355–366. *In* A. Berger et al. (ed.) Climate and geo-sciences. NATO ASI Ser. C. Vol. 285. Kluwer Academic Publ., Dordrecht, Netherlands.

Ehret, D.L., and P.A. Jolliffe. 1985. Photosynthetic carbon dioxide exchange of bean plants grown at elevated carbon dioxide concentration. Can. J. Bot. 63:2026–2030.

Enoch, H.Z., and B.A. Kimball. 1986. CO_2 enrichment of greenhouse crops. Vol. 1 and 2. CRC Press, Boca Raton, FL.

Farquhar, G.D., and S. von Caemmerer. 1983. Modelling of photosynthetic response to environmental conditions, encyclopedia of plant physiology. Vol. 12B. Springer-Verlag, New York.

Food and Agriculture Organization of the United Nations. 1987. FAO production yearbook. FAO, Rome, Italy.

Gaastra, P. 1962. Photosynthesis of leaves and field crops. Neth. J. Agric. Sci. 10:311–324.

Gezondheidsraad. 1983. Part One report on CO_2 problem (English translation). Recommendation to the Ministers of Public Health and of the Physical Environment issued by the Health Council of the Netherlands. Health Council, The Hague, Netherlands.

Gezondheidsraad. 1987. CO_2 problem, scientific opinions and impacts on society (English translation). Second advice and final report. Health Council, The Hague, Netherlands.

Gifford, R.M. 1979. Growth and yield of carbon dioxide-enriched wheat under water-limited conditions. Aust. J. Plant Physiol. 6:367–378.

Gifford, R.M., and J.I.L. Morison. 1985. Photosynthesis, water use and growth of a C_4-grass stand at high CO_2-concentration. Photosynth. Res. 7:77–90.

Gleick, P.H. 1986. Methods for evaluating the regional hydrologic impacts of global climatic changes. J. Hydrol. 88:97–116.

Gleick, P.H. 1987. Regional hydrologic consequences of increases in atmospheric CO_2 and other trace gases. Clim. Change 10:137–161.

Goudriaan, J. 1977. Crop micrometeorology: A simulation study. Simulation Monographs. Pudoc, Wageningen, Netherlands.

Goudriaan, J., and R.J. Bijlsma. 1987. Effect of CO_2-enrichment on growth of faba beans at two levels of water supply. Neth. J. Agric. Sci. 35:189–191.

Goudriaan, J., and H.E. de Ruiter. 1983. Plant growth in response to CO_2-enrichment, at two levels of nitrogen and phosphorus supply. Neth. J. Agric. Sci. 31:157–169.

Goudriaan, J., and H. van Keulen. 1979. The direct and indirect effects of nitrogen shortage on photosynthesis and transpiration in maize and sunflower. Neth. J. Agric. Sci. 27:227–234.

Goudriaan, J., and H.H. van Laar. 1978. Relations between leaf resistance, CO_2 concentration and CO_2 assimilation in maize, beans, lalang grass and sunflower. Photosynthetica 12:241–249.

Goudriaan, J., H.H. van Laar, H. van Keulen, and W. Louwerse. 1985. Photosynthesis, CO_2 and plant production. p. 107–122. *In* W. Day and R.K. Atkin (ed.) Wheat growth and modelling. NATO ASI Ser. A: Life Sciences. Vol. 86. Plenum Publishing, New York.

Jarvis, P.G., and K.G. McNaughton. 1986. Stomatal control of transpiration: Scaling up from leaf to region. p. 1–49. *In* Advances in ecological research. Vol. 15. Academic Press, New York.

Jones, P., L.H. Allen, K.J. Boote, and W.J. Campbell. 1984. Soybean canopy growth, photosynthesis, and transpiration responses to whole-season carbon dioxide enrichment. Agron. J. 76:633–637.

Kimball, B.A. 1983. Carbon dioxide and agricultural yield: An assemblage and analysis of 430 prior observations. Agron. J. 75:779–788.

Kimball, B.A., J.R. Mauney, G. Guinn, F.S. Nakayama, P.J. Pinter, K.L. Clawson, S.B. Idso, G.D. Butler, Jr., and J.W. Radin. 1984. Effects of increasing atmospheric CO_2 on the yield and water use of crops. *In* Response of vegetation to carbon dioxide. Ser. 023. U.S. Dep. of Energy, and USDA-ARS, Washington, DC.

Kimball, B.A., J.R. Mauney, G. Guinn, F.S. Nakayama, P.J. Pinter, K.L. Clawson, R.J. Reginato, and S.B. Idso. 1983. Effects of increasing atmospheric CO_2 on the yield and water use of crops. *In* Response of vegetation to carbon dioxide. Ser. 021. U.S. Dep. of Energy and USDA-ARS, Washington, DC.

Kimball, B.A., J.R. Mauney, J.W. Radin, F.S. Nakayama, S.B. Idso, D.L. Hendrix, D.H. Akey, S.G. Allen, M.G. Anderson, and W. Hartung. 1986. Effects of increasing atmospheric CO_2 on the growth, water relations, and physiology of plants grown under optimal and limiting levels of water and nitrogen. *In* Response of vegetation to carbon dioxide. Ser. 039. U.S. Dep. of Energy and USDA-ARS, Washington, DC.

Langbein, W.B., et al. 1949. Annual runoff in the United States. USGS Circ. 5. U.S. Dep. of the Interior, Washington, DC.

Lemon, E.R. (ed.). 1983. CO_2 and plants. AAAS Selected Symp. 84. Westview Press, Boulder, CO.

Lenssen, G.M. 1986. De invloed van CO_2 in combinatie met fosfaatgerbrek op de groei en de transpiratie van tarwe. M.Sc. thesis. Agricultural Univ., Wageningen, Netherlands (in Dutch).

Louwerse, W. 1980. Effects of CO_2 concentration and irradiance on the stomatal behaviour of maize, barley and sunflower plants in the field. Plant Cell Environ. 3:391–398.

Monteith, J.L. 1981. Climate variation and growth of crops. Q. J. R. Meterol. Soc. 107:749–774.

Morison, J.I.L. 1987. Intercellular CO_2 concentration and stomatal response to CO_2. *In* E. Zeiger et al. (ed.) Stomatal function. Stanford Univ. Press, Stanford, CA.

Mortensen, L.M. 1983. Growth responses of some greenhouse plants to environment: X. Long-term effect of CO_2 enrichment on photosynthesis, photorespiration, carbohydrate content and growth of *Chrysanthemum morifolium* Ramat. Sci. Rep. 272. Agric. Univ. of Norway. Vol. 62. The Agric. Univ. of Norway, Ås, Norway.

Nederhoff, E.M., and A.M. van Uffelen. 1988. Effects of continuous and intermittent carbon dioxide enrichment on fruit set and yield of sweet pepper (*Capsicum annuum* L.). Neth. J. Agric. Sci. 36:209–217.

Norby, R.J., E.G. O'Neill, and R.J. Luxmoore. 1986. Effects of atmospheric CO_2 enrichment on the growth and mineral nutrition of *Quercus alba* seedlings in nutrient-poor soil. Plant Physiol. 82:83–89.

Oldekop, E.M. 1911. On evaporation from the surface of river basins. Sbornik. trudov, ispolnevykh studentami pri Iurievskoj observatorii (Tartu). Univ. of Tartu, Estonia.

Palmer, T.N., and C. Brankovic. 1989. The 1988 U.S. drought linked to anomalous sea surface temperature. Nature (London) 339:54–57.

Parry, M.L., T.R. Carter, and N.T. Konijn. 1988. The impact of climatic variations on agriculture. Vol. 1: Assessments in cool temperate and cold regions. Vol. 2: Assessments in semi-arid regions. Kluwer, Dordrecht, Netherlands.

Pearcy, R.W., and J. Ehleringer. 1984. Comparative ecophysiology of C_3 and C_4 plants. Plant Cell Environ. 7:1–13.

Ramanathan, V., R.D. Cess, E.F. Harrison, P. Minnis, B.R. Barkstrom, E. Ahmad, and D. Hartmann. 1989. Cloud-radiative forcing and climate: Results from the Earth radiation budget experiment. Science (Washington, DC) 243:57–63.

Raschke, K. 1975. Stomatal action. Annu. Rev. Plant Physiol. 26:309–340.

Revelle, R.R., and P.E. Waggoner. 1983. Effects of a carbon dioxide-induced climatic change on water supplies in the Western United States. *In* Changing climate. Rep. of the Carbon Dioxide Assessment Committee. National Academy Press, Washington DC.

Rosenberg, N.J. 1981. The increasing CO_2-concentration in the atmosphere and its implication on agricultural productivity. Clim. Change 3:265–279.

Rosenberg, N.J., B.A. Kimball, Ph. Martin, and C.F. Cooper. 1990. From climate and CO_2 enrichment to evapotranspiration. p. 151–175 *In* P.E. Waggoner (ed.) Climate change and U.S. water resources. John Wiley & Sons, New York.

Rosenzweig, C. 1985. Potential CO_2-induced climate effects on North American wheat-producing regions. Clim. Change 7:367–389.

Schaake, J.C. 1990. From climate to flow. p. 176. *In* P.E. Waggoner (ed.) Climate change and U.S. water resources. John Wiley & Sons, New York.

Schlesinger, M.E., and J.F.B. Mitchell. 1985. Model projections of the equilibrium climatic response to increased carbon dioxide. *In* M.C. MacCracken and F.M. Luther (ed.) The potential climatic effects of increasing carbon dioxide. DOE/ER-0237. U.S. Dep. of Energy, Washington, DC. Available from NTIS, Springfield, VA.

Schwarz, M., and J. Gale. 1984. Growth response to salinity at high levels of carbon dioxide. J. Exp. Bot. 35:193–196.

Sionit, N., N. Hellmers, and B.R. Strain. 1980. Growth and yield of wheat under CO_2-enrichment and water stress. Crop Sci. 20:456–458.

Squire, G.R., and M.H. Unsworth. 1988. Effects of CO_2 and climatic change on agriculture, 1988. Rep. to the UK Dep. of Environ. Univ. of Nottingham, UK.

Strain, B.R., and J.D. Cure. 1985. Direct effects of increasing carbon dioxide on vegetation. DOE/ER-0238. U.S. Dep. of Energy, Washington, DC. Available from NTIS, Springfield, VA.

Thran, P., and S. Broekhuizen. 1965. Agro-ecological atlas of cereal growing in Europe. Vol. I. Agro-climatic atlas of Europe. Elsevier Science Publishing, New York.

Valle, R., J.W. Mishoe, W.J. Campbell, J.W. Jones, and L.H. Allen. 1985. Photosynthetic responses of 'Bragg' soybean leaves adapted to different CO_2 environments. Crop Sci. 25:333–339.

van Diepen, C.A., H. van Keulen, F.W.T. Penning de Vries, I.G.A.M. Noy, and J. Goudriaan. 1987. Simulated variability of wheat and rice yields in current weather conditions and in future weather when ambient CO_2 has doubled. Simulation Rep. CABO-TT Nr 14. CABO, Wageningen, Netherlands.

Wilks, D.S. 1988. Estimating the consequences of CO_2-induced climatic change on North American grain agriculture using General Circulation Model information. Clim. Change 13:19–42.

Wilson, C.A., and J.F.B. Mitchell. 1987. A doubled CO_2 climate sensitivity experiment with a global climate model including a simple ocean. J. Geophys. Res. 92:13 315–13 343.

Wittwer, S.H. 1980. Carbon dioxide and climatic change: An agricultural perspective. J. Soil Water Conserv. 35:116–120.

Wittwer, S.H., and W. Robb. 1964. Carbon dioxide enrichment of greenhouse atmospheres for food crop production. Econ. Bot. 18:34–56.

Wong, S.C. 1979. Elevated atmospheric partial pressure of CO_2 and plant growth. I. Interactions of nitrogen nutrition and photosynthetic capacity in C_3 and C_4 plants. Oecologia 44:68–74.

APPENDIX A

List of Volunteered Poster Presentations at the Symposium on Impact of Carbon Dioxide, Trace Gases, and Climate Change on Global Agriculture, 1 Dec. 1988, Anaheim, CA

Titles, authors, and institutions	Page†
Growth and Yield of Rice as Affected by Carbon Dioxide Concentration L.H. Allen, Jr.,‡ J.T. Baker, K.J. Boote, J.W. Jones, and P. Jones. USDA-ARS, University of Florida, Gainesville, FL.	12
Herbivorous Insect Responses to Cotton Grown in Enriched Carbon Dioxide Atmospheres D.H. Akey,‡ B.A. Kimball, and J.R. Mauney. USDA-ARS, Phoenix, AZ.	12
Interactive Effects of CO_2 and Environment on Photosynthesis of Two Aquatic Plant Species S.G. Allen,‡ S.B. Idso, and B.A. Kimball. USDA-ARS, Phoenix, AZ.	12
Mode of Action of ABA at Enriched CO_2 Environment on Rooting Cuttings of *I. balsamina* in Relation to Photosynthate S. Bhattacharya,‡ B.R. Strain, A.W. Naylor, and N. Bhattacharya. Duke University, Durham, NC.	13
Net Assimilation Rate and Relative Growth Rate of Sweet Potatoes in Response to Enriched CO_2 under Field Conditions N.Bhattacharya,‡ P.P. Ghosh, D.R. Hileman, and P.K. Biswas. Tuskegee University, Tuskegee, AL.	13
Responses to Rice to Photoperiod and Carbon Dioxide Concentration J.T. Baker,‡ L.H. Allen, Jr., K.J. Boote, J.W. Jones, and P. Jones. University of Florida, USDA-ARS, Gainesville, FL.	13
Changes in Soybean Leaf Growth, Photosynthesis, and Canopy Structure during Adjustment to Doubled CO_2 J.D. Cure,‡ T.W. Rufty, Jr., and D.W. Israel. Duke University, USDA-ARS, Raleigh, NC.	16
Climate Change, Carbon Dioxide Enrichment, and Evapotranspiration: Wheat, Forest, and Grass P.H. Martin‡, N.J. Rosenberg, and B.A. Kimball. University of California, Berkeley, CA; Resources for the Future, Washington, DC; and U.S. Water Conservation Lab., Phoenix, AZ.	23
Potential Effects of Climate Change on Wheat and Corn Production in the Great Plains C.E. Rosenzweig‡. Columbia University, NASA-Goddard Inst. for Space Studies, New York, NY.	26

(continued)

† Page number in *Agronomy Abstracts*, 1988, American Society of Agronomy, 677 S. Segoe Road, Madison, WI 53711.

‡ Indicates presenter.

Appendix A. Continued.

Titles, authors, and institutions	Page†
Rice Stomatal Density, Photosynthetic Rate, and Rubisco Activity as Affected by Carbon Dioxide Concentration A. Rowland-Bamford‡, L.H. Allen, Jr., J.T. Baker, K.J. Boote, and G. Bowes. University of Florida, USDA-ARS, Gainesville, FL.	26
Effects of Long-Term Exposure to Elevated Atmospheric CO_2 Concentration on Carbon Exchange Rates of Two Perennial Plants N. Sionit,‡ N. Fletcher, C.H. Jaeger, and B.R. Strain. San Diego State University, San Diego, CA, and Duke University, Durham, NC	48
Responses of Field-Grown Cotton to Enrichment with Carbon Dioxide J.R. Mauney,‡ B.A. Kimball, and J.W. Radin. USDA-ARS, Phoenix, AZ.	114
CO_2-Enriched Photosynthesis as Affected by a Rapidly Induced Phosphorus Stress W.T. Pettigrew,‡ J.D. Hesketh, and D.B. Peters. University of Illinois, USDA-ARS, Gainesville, FL.	115
Water Use Efficiency of Sorghum and Winter Wheat as Affected by Enriched CO_2 U.N. Chaudhuri,‡ E.T. Kanemasu, and M.B. Kirkham, Kansas State University, Manhattan, KS.	123

† Page number in *Agronomy Abstracts*, 1988, American Society of Agronomy, 677 S. Segoe Road, Madison, WI 53711.

‡ Indicates presenter.

APPENDIX B

Units for Carbon Dioxide Concentration and Emissions

To convert from: [To convert to:]	To: [From:]	Multiply by: [Divide by:]
	CO_2 concentration†	
ppm or ppmv or vpm (parts per million on a volume basis)	μmol CO_2 mol^{-1}	1.0
μL CO_2 L^{-1} or cm^3 CO_2 m^{-3} (where the denominator refers to a liter or m^3 of air at the same temperature and pressure as the CO_2 and not to a liter or m^3 of physical space)	μmol CO_2 mol^{-1}	1.0
μbar CO_2 (partial pressure)	μmol CO_2 mol^{-1}	1.01325‡
Pa CO_2 (partial pressure)	μmol CO_2 mol^{-1}	10.1325§
mg CO_2 m^{-3} (density at STP¶)	μmol CO_2 mol^{-1}	0.5092¶
μmol CO_2 m^{-3} (molar density at STP¶)	μmol CO_2 mol^{-1}	0.02241¶
ppm (parts per million on a weight basis)	μmol CO_2 mol^{-1}	0.6582
	Weight of carbon (or anything)	
g (gram)	Pg (petagram)	10^{-15}
gigaton	Pg	1.0
g (gram)	Tg (teragram)	10^{-12}
	CO_2 Emissions	
μg CO_2 m^{-2} s^{-1}	Pg CO_2 yr^{-1} global area^{-1}	16.1#
μmol CO_2 m^{-2} s^{-1}	Pg CO_2 yr^{-1} global area^{-1}	710#
Pg C yr^{-1} global area^{-1} (carbon in CO_2)	Pg CO_2 yr^{-1} global area^{-1}	3.664

† The preferred units are μmole of CO_2 per mole of air where the "air" is usually implied.

‡ At standard sea level barometric pressure of 1.01325 bar. Divide μbar CO_2 by actual barometric pressure in bar for exact conversion.

§ At standard sea level barometric pressure of 101.325 kPa. Multiply Pa CO_2 by 1000 and divide Pa CO_2 by actual barometric pressure in kPa for exact conversion.

¶ At standard temperature of 273.16 °K and pressure of 101.325 kPa where the m^3 refers to a volume of physical space. Additionally, multiply by (T/273.16 °K) (101.352 kPa/P) for other temperature, T, and pressure, P.

Based on 6380 km for the radius of a spherical Earth.